George Edward Mannering

With Axe and Rope in the New Zealand Alps

George Edward Mannering

With Axe and Rope in the New Zealand Alps

ISBN/EAN: 9783743418837

Manufactured in Europe, USA, Canada, Australia, Japa

Cover: Foto ©Andreas Hilbeck / pixelio.de

Manufactured and distributed by brebook publishing software (www.brebook.com)

George Edward Mannering

With Axe and Rope in the New Zealand Alps

WITH AXE AND ROPE

IN

THE NEW ZEALAND ALPS

BY

GEORGE EDWARD MANNERING

MEMBER OF THE NEW ZEALAND ALPINE CLUB
MEMBER OF THE ROYAL GEOGRAPHICAL SOCIETY OF AUSTRALASIA
MEMBER OF THE PHILOSOPHICAL INSTITUTE OF CANTERBURY, N.Z.

WITH ILLUSTRATIONS

LONDON
LONGMANS, GREEN, AND CO.
AND NEW YORK: 15 EAST 16th STREET
1891

All rights reserved

THIS BOOK IS DEDICATED
TO ALL LOVERS OF NATURE

807277

PREFACE

This short work contains the story of five seasons' climbing and exploring in the New Zealand Alps. Most of the material embodied in it has already appeared from time to time, in rather a different form, in the Christchurch (N.Z.) 'Weekly Press.'

The author trusts that the publication of the same in book form, together with a map of the locality and a few photographic reproductions, will supply a want in the shape of a guide-book to the Alpine mountain district which is already beginning to be felt by tourists in New Zealand; and he hopes that the contents may not prove uninteresting to the general public, more especially to Swiss and Caucasian climbers, few of whom are perhaps aware of the extent and nature of the New Zealand Alpine chain.

The map is compiled by the New Zealand Government Survey Office from the work of Mr. T. N. Brodrick, Government Surveyor, and that of Dr. R. von Lendenfeld. The illustrations are from photographs by Messrs. Wheeler and Son. Their operator has in several mountain expeditions accompanied the author, who takes this opportunity of expressing his thanks to the New Zealand Government Survey Department, and to Messrs. Wheeler, for their kind assistance.

It will doubtless be said that the summit of Aorangi has not yet been attained: quite true. Like Mr. Green, the author and his friend were 'wise in time.' Yet it is only a quibble to dispute the ascent of the mountain, for being on the ice-cap of Aorangi is like being on the topmost rung of a ladder, and yet not upon the projections above that step.

CHRISTCHURCH, NEW ZEALAND:
April 13, 1891.

CONTENTS

CHAPTER I
INTRODUCTORY

	PAGE
The New Zealand Alps and their glaciers	1

CHAPTER II
THE ROUTE TO THE MOUNT COOK DISTRICT

A short description of the route to the Mount Cook district, and of the topographical features of the Mueller, Hooker, and Tasman Valleys. 5

CHAPTER III
FIRST ATTEMPT TO CLIMB AORANGI

First impressions—Swagging—The Hochstetter Glacier—Defeat—The perils of river-crossing 11

CHAPTER IV
SECOND ATTEMPT TO CLIMB AORANGI

A flooded camp in the Tasman Valley—Hard struggles—We reach Green's bivouac 32

CHAPTER V
THIRD ATTEMPT TO CLIMB AORANGI

Photography on the Tasman Glacier—Attempt to scale Mount De la Bêche 42

CHAPTER VI
ASCENT OF THE HOCHSTETTER DOME

Camp under De la Bêche—Twelve hours on snow and ice—The pangs of hunger 58

THE NEW ZEALAND ALPS

CHAPTER VII

FOURTH ATTEMPT TO CLIMB AORANGI

We reach the Great Plateau at last—Defeat again—The crossing of the Ball Pass 65

CHAPTER VIII

FIRST EXPLORATION OF THE MURCHISON GLACIER

Hard swagging—Erroneous maps—The struggle for Starvation Saddle—Exhaustion and hunger—Return 76

CHAPTER IX

FIFTH ATTEMPT TO CLIMB AORANGI

Avalanches—The bivouac again—First attempt repulsed—Second attempt—The Great Plateau—The Linda Glacier—Hard work step-cutting—The terrible *couloirs*—Victory at last—Descent by lantern-light—Back to civilisation 90

CHAPTER X

ON SOME OF THE PHENOMENA OF GLACIERS

The cause of glaciers—Formation and structure—Motion—Moraines: Lateral, medial, and terminal—'Surface' moraines—Crevasses—Moulins—Glacier tables—Glacier cones—Surface torrents—Avalanches—Cornices 109

CHAPTER XI

CANOEING ON THE NEW ZEALAND RIVERS

The Waimakariri—The enormous rainfall—Descent of the Waitaki River—The Tasman branch—Lake Pukaki—Leaky canoes—The Pukaki Rapids—The Waitaki Gorge—Out on the plains again—Sixty miles paddle to catch the train—Home once more . 119

L'Envoi 131

Appendix 133

A Short Glossary of Technical Alpine Terms 139

LIST OF ILLUSTRATIONS

CLIMBING PARTY ON THE TASMAN GLACIER	*Frontispiece*
HOOKER VALLEY AND MOUNT SEFTON FROM GOVERNOR'S CAMP	*To face page* 8
AORANGI: MOUNT COOK AND THE HOOKER GLACIER	,, 10
MOUNT TASMAN (11,475 FEET) AND HOCHSTETTER ICE-FALL	,, 28
MOUNT COOK AND THE HERMITAGE	,, 46
CROSSING THE HOOKER RIVER	,, 48
AORANGI FROM THE BALL GLACIER	,, 50
ICE CAVE, TASMAN GLACIER	,, 52
MOUNT DE LA BÊCHE (10,021 FEET) FROM THE TASMAN GLACIER	,, 54
PEAKS ON MALTE BRUN	,, 58
THE TASMAN GLACIER	,, 66
MOUNTAIN LILIES (*Ranunculus Lyalii*)	,, 86
LOOKING ACROSS THE MURCHISON GLACIER	,, 90
AORANGI FROM THE TASMAN GLACIER	,, 90
THE MURCHISON GLACIER	,, 92
AORANGI: THE HIGHEST PEAK	,, 100
IN THE ICE-FALL OF THE ONSLOW GLACIER	,, 120
THE SURFACE OF A GLACIER	,, 128
MAP	*At end*

WITH AXE AND ROPE
IN
THE NEW ZEALAND ALPS

CHAPTER I

INTRODUCTORY

The New Zealand Alps and their glaciers

It is unnecessary for me in these days of universal education and enlightenment to describe the geographical position of New Zealand, the 'Britain of the South,' and the future playground of Australasia.

Everyone knows that New Zealand consists of three islands, situate between the 34th and 47th degrees of south latitude, off the south-east coast of Australia. Reference to almost any handbook of the colony will furnish every information regarding settlement, population, government, climate, and so on, and I do not propose to dwell longer than is necessary on any general matters of this nature.

It is advisable, however, to describe in as brief and concise a manner as possible the general physical features of a country containing such varieties of scenery and climate, more especially those of the South

Island (or Middle Island as it is sometimes called), where the High Alps and their wonderful glaciers are situated.

Speaking generally, the highest mountains of New Zealand may be said to run in a north-easterly direction from the southernmost point of the South Island through the whole country, like a vast backbone, to the north-eastern point of the North Island.

The main formation of the mountains dates back to Jurassic times, so that the geological structure may be said to be one of great antiquity.

Volcanic action has long since ceased throughout the South Island; but there are many active volcanoes in the North, where a perfect wonderland of hot-springs, solfataras, and silica terraces exists.

As a whole, the country is heavily timbered—more thickly on the western parts, where the greater rainfall occurs. This is notably the case in the South Island, where the hot and moisture laden winds from tropical regions are deprived by the Alpine chain of their aqueous vapour.

The Southern Alps proper may be said to extend over a distance of about one hundred miles of the middle part of the South Island, the chain being situated closer to the western than the eastern ocean. The slopes on the western side are the more precipitous, and are clothed with heavy timber and intersected by innumerable mountain torrents, fed in most cases by glaciers, some of which descend to within 600 feet of sea-level.

Ranges of outlying foot-hills occur on the eastern side, snow-covered in winter, amongst which many large glacier-fed rivers have cut their way, and meander over the plains (probably of entirely fluviatile formation) which slope gradually from the outer bases of the foot-hills to the eastern ocean.

The peaks of the Alps range in height from 7,000 to 12,350 feet above sea-level, the majority of those over 10,000 feet being contiguous to the culminating point in altitude—Aorangi—more popularly known as Mount Cook. Here also are found the largest glaciers.

The snow-line is a low one when compared with that of Alpine countries in the northern hemisphere and in relative latitudes. It would be difficult to compute its average altitude, but in parts where large glaciers and snow-fields exist it is even as low as 5,000 feet above sea-level.

By comparison with Switzerland, for instance, it may safely be said that the snow-line in New Zealand is from 2,000 to 3,000 feet lower; consequently we have the same Alpine conditions at a much lower level. Owing to this interesting fact, we find that the New Zealand glaciers attain far greater dimensions than those of Switzerland, although the peaks do not rise to such a height above sea-level.

In themselves, I believe the mountains compare favourably as to size or actual height above the valleys below them; Aorangi, for instance, rising for nearly 10,000 feet from the Hooker Glacier, and Mount Sefton 8,500 feet from the Mueller Glacier, whilst the western precipices of Mount Tasman (11,475 feet) are stupendous.

The enormous length attained in remote times by the New Zealand glaciers is evident on all hands at the lower parts of the valleys, the heads of which they now occupy; whilst the formation of nearly all the lakes in the South Island can be traced to the action of ice and the deposition of terminal moraines, prior to a period of retreat of the ice.

There is an interesting feature in the glaciers of this country peculiar to them; I refer to the deposition of singularly extensive moraines. The lower parts of the large glaciers on the eastern slopes are, in nearly every instance, completely covered with accumulated *débris* derived from the moraines. This is variously accounted for by the antiquity of the mountain chain, the slow rate of motion in the ice, and great denudation from rocks which are much jointed and offer but little resistance to the splitting powers of freezing infiltrated water.

The western glaciers I am not personally acquainted with, but I understand that they do not carry anything like the amount of moraine, and I imagine the cause of the disparity will be found in a faster motion of the ice, and (a yet more potent factor) in the dip of the strata of the rocks, which is from east to west, the broken faces being eastward and the slab-like faces westward.

CHAPTER II

THE ROUTE TO THE MOUNT COOK DISTRICT

A short description of the route to the Mount Cook district, and of the topographical features of the Mueller, Hooker, and Tasman Valleys

From Timaru on the east coast the traveller may comfortably reach the glaciers of Aorangi in a two days' journey.

Leaving Timaru by an evening train, Fairlie Creek (the present terminus of the railway line) is reached, where the night is spent. Two days' coaching then are required to cross over Burke's Pass into the great Mackenzie plains, across this great ancient glacier bed, past Lakes Tekapo and Pukaki, over the rivers of the same names, and up the valley of the Tasman River to a comfortable hostelry called 'The Hermitage,' nestling right under the shadow of that wonderful pile of ice-clad mountain glory, Mount Sefton.

Lakes Tekapo and Pukaki may both be aptly compared in one way to the Lake of Geneva, in that they are of glacier origin, and purify the rivers which now flow from the present glaciers, parting with their waters again through channels cut in the ancient terminal moraines which dam their respective southern shores.

They are both beautiful, each in its own way—

Tekapo sunny, peaceful, and calm; Pukaki awe-inspiring and grand—but they lack the charm of chalet and pine tree, of vine and meadow, which so adorn the shores of the Swiss lakes.

The immediate vicinity of the road is uninteresting, except from a geological point of view, for it winds about amongst old moraines, whose vegetation consists almost entirely of the brown tussock grass so general in the South Island.

Yet the geologist or student of glacier phenomena can read on the surface the history of the formation; *roches moutonnées* abound, and, in places, old moraines are spread over the bed rock for miles together, whilst erratic blocks are dotted about in various directions, evidencing how extensive has been the action of the ice in ages gone by.

Though the scenes contiguous to the road may fail to charm the eye, the distant panoramas of the glorious Southern Alps cannot fail to draw forth expressions of wonder from the most callous observer. As the Hermitage is approached, and the great peaks and glaciers draw closer and closer, the marvellous grandeur of the chain is gradually realised.

The sight of the reflection of Aorangi in Lake Tekapo, on a calm morning, is something to remember for a lifetime. The subject has long been a favourite one for brush and pen, but no one yet has done it justice.

A substantial bridge spans the exit of the Tekapo River, but only a ferry stage exists at the Pukaki River where it leaves the lake. A wire rope, 450 feet long, is thrown across the stream, to which the ferry stage floating on two punts is attached by runners. The coach and four is driven bodily on to the stage, and by the aid of a rudder the punts are slued so as to point across the stream diagonally. The force of the water rushing obliquely on to the sides of the punts drives the whole affair across in a space of about three or four minutes. This ingenious plan is commonly adopted in the New Zealand rivers.

During the months of winter it is possible to reach the Hermitage direct from Tekapo, and thus avoid striking south to go round Lake Pukaki, by crossing the Tasman River. During summer, however, as a rule, this river is impassable, for it rises so fast during warm and nor'-west weather from rain and melting snow that sometimes the whole bed of the river—two miles wide—is a network of rushing yellow torrents quite unfordable by man or beast.

Readers of the Rev. W. S. Green's 'High Alps of New Zealand' will recollect that his conveyance found a last resting-place in the quicksands of the Tasman. Von Lendenfeld also, the year after Mr. Green, experienced an unhappy week's delay on the eastern bank of the river. I have myself narrowly escaped drowning at the same point, and in years gone by the Tasman River has been accountable for more than one life.

The river in full flood is a sight to see; the water in places runs fifteen knots an hour, or even more. In the rapids it is piled up in the middle from sudden contraction of the banks, and forms crested billows four or five feet in height, whilst now and then a block of ice from the glacier may be seen bowling along.

The ancient glacier-formed terraces of the Tasman

Valley are instructive and interesting. The highest of them are distinctly marked all down the valley for a distance of forty miles from Sebastopol—a large face of ice-worn rock near the Hermitage—on the eastern slopes of the Ben Ohau Range. The story of the ancient glacier can be read as the eye follows these strange terraces from their starting point 2,000 feet above the valley bed, down a gentle declination to the terminus of the Ben Ohau Range.

Before going into the narrative of my five seasons' climbing amongst the peaks and glaciers around Aorangi, it would be as well for me to describe, as concisely as possible, the general topography of the Mueller, Hooker, and Tasman Glaciers.

We will suppose ourselves in the main Tasman Valley, into which all these glaciers drain, close to the point where the valley first branches. As we look northward, Aorangi and the range running southward for twelve miles from the main body of the mountain bound the view, and divide the valley into two branches. Let us take the one to the north-west first. Proceeding up this valley of the Hooker for a few miles, we arrive at a branch valley from the left or west—the Mueller Valley—completely occupied by the glacier of the same name. Close to the Mueller Glacier is situated the Hermitage, presided over by Mr. F. F. C. Huddleston, a true haven of refuge and comfort for the wearied tourist or mountaineer.

We step on to the Mueller Glacier, here completely covered by moraine, and, turning westwards, strike up its course. On our right, 8,500 feet above us, clad in ice and snow and glittering in the sunlight, rises the

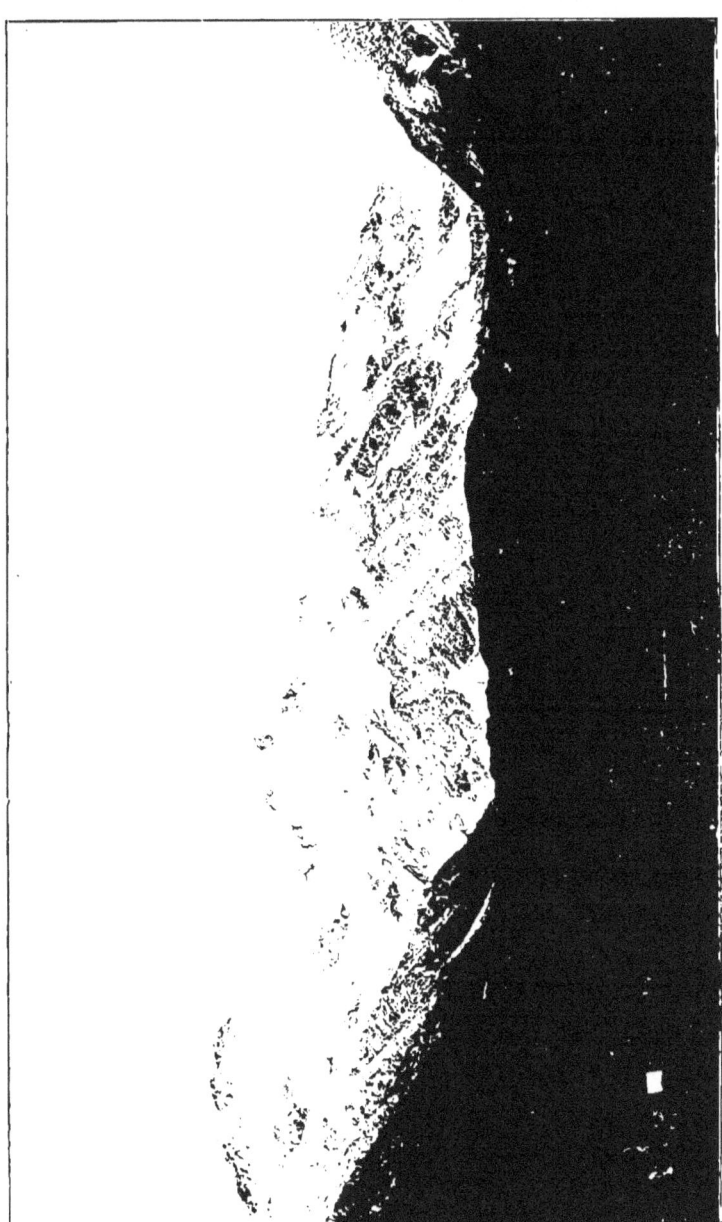

HOOKER VALLEY AND MOUNT SEFTON FROM GOVERNOR'S CAMP

[Wheeler & Son, Photo.

glorious mass of Mount Sefton, showering down avalanches upon the glacier. On our left the shingle slips from the rotten and crumbling crags of the Sealy Range. It is possible for tourists who are good walkers to reach the head of this glacier, which is seven and a half miles long and about one mile broad, in one day. The moraine gives way to the clear ice some three miles or so from the terminal face. Now we return and make a fresh start up the Hooker Valley due northwards from the Hermitage.

Crossing the Mueller Glacier we walk through a perfect garden of lilies (*ranunculus Lyallii*), celmisias, 'Spaniards,' and an endless variety of sub-alpine plants, for a distance of about one mile from the northern side of the Mueller Glacier, when we come to the terminal moraine-covered face of the Hooker Glacier.

On our right rises up the bold and verdure clad snow-topped Mount Cook Range, Mount Wakefield (6,561 feet), Mount Mabel (6,868 feet), Mount Rosa (6,987 feet), and a nameless peak (7,540 feet) being the principal points of interest. On our left is the northern continuation of the ridge of Mount Sefton, known as the Moorhouse Range, part of the main chain of the Southern Alps. Several secondary glaciers descend from the slopes, but do not reach the bed of the valley below, which is filled from side to side with the Hooker Glacier.

Proceeding up the surface of the glacier we get on to the clear ice, and now on either bank the mountains rise to a great height. On the right Aorangi suddenly rears itself, from a point known as the Ball Saddle (7,500 feet), to 12,349 feet in one stupendous rocky ridge, upon which the ice hangs wherever it can get any hold. This ridge is known to climbers as the Great Southern *arête*, and has been found, first by Mr. Green and secondly by myself, to be inaccessible. Right ahead of us pour down from the highest crags the Mona, Noeline, and Empress Glaciers, to join the Hooker, alternating with very precipitous rocky ridges which present every appearance of being quite unscalable.

Several attempts have been made by surveyors and others to reach the saddle at the head of the Hooker, but it was only in December 1890 that the efforts of two climbers (Mr. A. P. Harper and Mr. R. Blakiston) were rewarded. The expedition can only be attempted with any chance of success in the early part of the season, when the numberless crevasses are yet covered with the winter snow.

From the Hooker Glacier we turn our faces downwards to the south again, and pay a visit to the northeastern branch of the main Tasman Valley.

Crossing the Hooker River at the terminal point of the Mount Cook Range, where a cage swung on a wire rope over the river now facilitates the traveller's passage, we strike north-eastwards up the valley.

For a distance of four miles our way leads over the shingle and boulder flats of the Tasman river-bed, here some two miles wide. Patches of good sheep-feed consisting of tussock and cocksfoot grass (the latter sown by an early settler) occur on the western side of the valley, but the river as a rule washes the opposite slopes.

Arriving at the terminal face of the glacier we

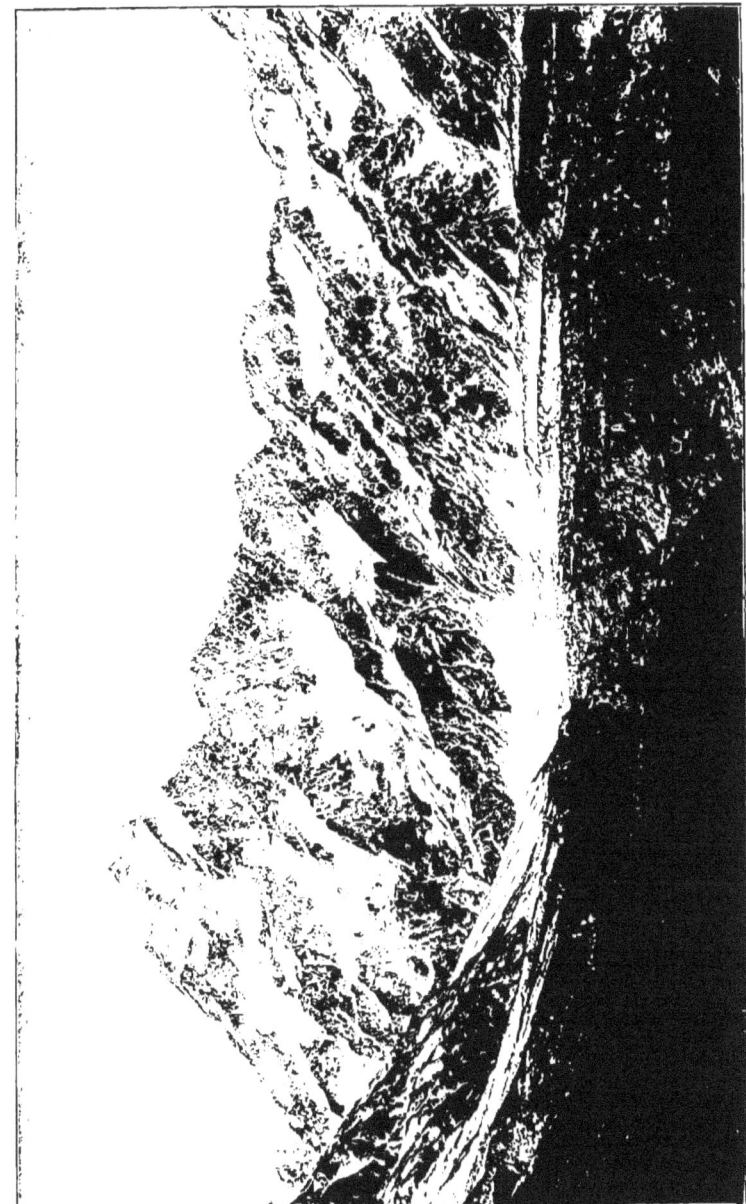

AORANGI: MOUNT COOK AND THE HOOKER GLACIER.

strike up a small valley between the western lateral slopes of the moraine of the glacier and the Mount Cook Range on our left, and for a distance of about seven or eight miles force our way through dense scrub and loose boulders from the moraine and mountain slopes, to the junction of the Ball Glacier with the Tasman. This Ball Glacier comes from the Great Southern *arête* of Aorangi, and is fed almost entirely by avalanches, there being no snow-fields—or *névés* as they are called in Alpine parlance—of any great extent at its head.

From this point upwards we strike out on to the ice on our right, and another seven miles or so brings us to a further division of the valley, Mount de la Bêche being the dividing peak. The glacier of the left-hand or northern branch is known as the Rudolf Glacier, whilst the main body of the Tasman stretches some six miles further north-eastwards to the Hochstetter Dome, where it again divides. The saddle at the head of the left-hand branch, again, has been reached by Dr. von Lendenfeld and by myself in our respective ascents of the Hochstetter Dome, and commands a superb view of the Whymper Glacier and valley, and of the Wataroa River on the west coast. The head of the branch to the right of the Hochstetter Dome has not yet been reached by man.

Taking a retrospective glance again at the peaks on either hand, and commencing at the lower end of the glacier, we have first on our right the Liebig Range till opposite the Ball Glacier, when the *embouchure* of the Murchison Valley occurs, followed by the Malte Brun Range, with the main peak—the Matterhorn of New

Zealand—opposite to Mount de la Bêche, then the Darwin Glacier followed by the mountain of the same name, and then the saddle between Mount Darwin and the Hochstetter Dome.

Now, again, on the left or western side of the great glacier we have the Mount Cook Range for ten miles, the Ball Glacier, Aorangi, the Hochstetter Glacier, Mounts Tasman, Haast, Haidinger, Glacier Peak, Mounts Spencer, Kant, Rudolf (at the head of the Rudolf Glacier), De la Bêche, Green, and Elie de Beaumont, the last followed by the Lendenfeld Saddle, to which I have already referred.

From Mount Tasman northwards to this saddle all these mountains are situated in the main chain. Aorangi itself, though popularly believed to belong to the main divide, is in reality separated from it by a rocky ridge and a saddle of about 10,500 feet, which leads to the Hooker Glacier on the one hand and the Linda on the other, both being east of the main divide. Aorangi itself, therefore, consists of a divergent ridge, the whole of whose drainage goes eastward.

Though for some years I have believed this to be the case, it is only quite recently that I have been able to substantiate the belief by ocular demonstration, when the ascent of the mountain was accomplished by Mr. Dixon and myself. To this expedition I shall refer later on.

The reader must picture to himself the great Tasman Glacier, nearly two miles in width and eighteen to twenty in length, occupying the whole of the bed of the valley, and fed on both sides by numerous tributary ice-streams from the mountains.

Of the Murchison Valley it is not necessary for me to speak just now, as the topographical features will be described when I come to tell the story of its exploration. Neither is it needful to refer in further detail to the Tasman for the same reason.

CHAPTER III

FIRST ATTEMPT TO CLIMB AORANGI

First impressions—Swagging—The Hochstetter Glacier—Defeat - The perils of river crossing

'To climb steep hills requires slow pace at first.'

IT was on March 24, 1886, that I left Christchurch, in company with my cousin, Mr. C. D. Fox, on my first visit to the great Tasman Glacier and Mount Cook, or Aorangi.[1]

I often look back now with feelings of amusement at the audacity with which we determined to make our first attempt to scale the great monarch of the Southern

---

[1] The Maori name of Mount Cook is 'Aorangi,' or, more properly, 'Ao-Rangi.' The commonly accepted meaning of the term is 'Sky-piercer,' but as the Maori language admits of many varieties of translation, each version hovering about the region of true meaning, it is only natural that authorities should differ as to the correct construing of the word.

One good Maori scholar, whose reputation as such is almost pre-eminent, gives the poetical translation of 'Light of Day'—a singularly beautiful one, for it is the first peak to catch the morning light and the last to show the glow of evening.

Another very well-known Maori scholar, the Rev. J. W. Stack, assures me that the most reasonable interpretation that can be put upon the word 'Ao-Rangi' is 'Sent Peak'; and this is a singularly apt one, for the prevailing nor' west winds always cause condensation and the gathering of cloud-banners about the higher parts of the mountain. 'Heaven-piercer' and 'Cloud-piercer' are also often used, but are to a certain extent fancy names.

Alps, and wonder how we could have been so self-satisfied with our own powers and confident of our ability in undertaking such a gigantic task. I can only suppose that it was ignorance of what lay before us, and a clear case of 'fools rush in where angels fear to tread'; for when my thoughts run back over the toils, hardships, and bitter lessons of experience undergone during the past six years, and when I think of the position of two completely inexperienced men (as far as *true* Alpine work is concerned) launching straight out into such an undertaking, my heart seems to quail at the idea. It is true that we both had heard and read of much Alpine work, and had been for some time in touch with climbing-men, also we were both practised in hill-walking and accustomed to such work as mustering sheep, pig-hunting, and shooting over what in England would be termed rough mountains, so that as cragsmen we could scarcely be classed as novices. As to any knowledge other than theoretical of the conditions of snow and ice, however, we might be termed tyros, though Fox had done a little scrambling on the Swiss glaciers. Nevertheless, we had sufficient 'cheek' to consider ourselves wise and strong enough to go straight into a really difficult piece of Alpine work, and, laughing at all discouragement, we set off for the mountains.

I have already described the customary route to the glaciers of Mount Cook, so will not weary my readers with a long narrative of the journey.

At Timaru (four hours by rail from Christchurch) we completed our stock of provisions, consisting of biscuits, tinned meats, &c., and took the evening train on to Fairlie Creek (forty miles further inland), where on arrival we hired a horse and buggy and drove to Ashwick Station, seven miles distant on the road to the mountains.

The next day's journey took us over Burke's Pass and into the Mackenzie country, past the beautiful Lake Tekapo, and on to the ferry situate at the southern end of Lake Pukaki.

The road itself winds through bleak tussock plains, interesting only from a geological point of view; but all monotony of the immediate surroundings is completely lost when one looks further afield and gazes on the marvellous beauty of such scenes as the Southern Alps from Lake Tekapo, or the Ben Ohau Range from the plains. Even the most fastidious globe-trotter could not fail to be deeply impressed with such a picture as Aorangi from Lake Pukaki.

To look at Aorangi from this approach is enough to damp the spirit of the stoutest Alpine climber that ever breathed, and is quite sufficient to account for the disbelief and incredulity cherished in the mind of many a shepherd in the Mackenzie country regarding the possibility of ascending the peak.

History repeats itself, and just as we hear of the native mountaineers of the Himalayas, Andes, and Caucasus discrediting ascents of glacier peaks around whose very bases they and their ancestors have lived and died, so we find that our own countrymen, whose calling needs their constant presence amongst their flocks on the lower ranges, refuse to believe that mountains presenting such an appearance as Aorangi are in any manner of way to be scaled.

The following day brought us to the Hermitage. A low mist had hidden the higher peaks throughout the day, and led to a surprise on the following morning which I little dreamt of.

I wonder if all Alpine climbers, in first 'tasting the sweets of climbing,' are similarly impressed with their initial Alpine view!

No words of mine can describe the ecstasy which seemed to pervade my whole being as on the early, cloudless morning the wonderful picture of Mount Sefton reared itself in indescribable sunlit grandeur above the old bush-clad moraine close by the Hermitage. Here, indeed, was a new and a fairy-like world to live in. As we sat in the verandah of the Hermitage the ice-seamed crags appeared to rise up and up until they culminated in a long serrated and corniced ridge, seeming almost to overhang the very spot where we rested.

A scene of mountain glory never to be forgotten, a memory to last a lifetime!

More than 8,000 feet above us were built up those ice-clad precipices, their glaciers glinting in the bright morning light, their avalanches tearing down the mountain sides and waking the echoes of a hundred ravines and valleys with their thunder.

Where is the man who can describe these

> palaces of Nature, whose vast walls
> Have pinnacled in clouds their snowy scalps?

Where is the mountaineer—not the mere gymnast, but the Nature-loving mountaineer—who can tell the feelings of such a first impression?

And yet even this scene seems to fade in the memory and suffer by contrast with those of other pictures in the New Zealand Alps, for up the Tasman Valley, where later on in the day we wended our way, fresh vistas of Alpine glory were unfolded to view.

Aorangi from the Hermitage is also a grand sight. The mountain seems to possess a startling individuality and a majestic grandeur somewhat different in character from its worthy neighbour Mount Sefton. The view is more distant, but the bold outline of the peak stands out in relief against the blue of the heavens, and rears a face of glacier-clad precipices to a height of 10,000 feet above the Hooker Valley at the mountain foot. Light clouds float about the peak and lend an ethereal air to its beauty, imparting a fairy-like, floating appearance to the peak itself. At other times the outlines are apparently clear cut against the sky, giving an air of lasting and monumental dignity, and conveying the idea of stability from past ages to ages to come.

After an early lunch, and accompanied by Mr. Huddleston (the landlord of the Hermitage), and one of his men, we started off for the Tasman Glacier. The first part of the way leads down over stony flats to the termination of the Mount Cook Range, and at this point the Hooker River is crossed.

On this occasion we double-banked over on horseback without much difficulty; but very often the Hooker River is quite impassable with horses, the torrent being confined in a narrow boulder bed of about 200 feet in width, which in flood time, during the warmer months of spring and summer, is quite filled with a roaring torrent, often bearing down with it blocks of ice from the Mueller and Hooker Glaciers above.

Turning in a north-easterly direction round the end of the range we shaped our course up the Tasman Valley, and in two hours' time from the Hermitage arrived at the terminal face of the great glacier, which fills the whole of the valley from side to side, a width of about two miles. Here, then, the hard work was about to begin, for the horses could not proceed further, and it was necessary to carry everything from this point on our own backs.

Ah! good reader, have you ever carried a swag, a *real* swag—not a Swiss knapsack—but a real, torturing, colonial swag? When you take it up and sling it on your back in the orthodox fashion you remark: 'Yes; I think it *does* weigh fifty pounds.' In ten minutes your estimate of its weight has doubled. In an hour you begin to wonder why Nature has been so foolish as to make men who will carry swags; bad language seems to slip out 'quite in a casual way,' and you begin to bend forward and do the 'lift.' But the 'lift' does not seem to fulfil quite all that is said in its praise, for soon the torturing burden settles down again and drags on to your shoulders more heavily than ever. After a bit of nice balancing over loose moraine the swag triumphs. Down you go, and the wretched thing worries you, whilst you bark your fingers and swear horribly, bruising your knees and shins, and cursing the day on which you saw the light of a hard and feelingless world. You recover and repeat the performance as before, and by the time your day's work is done you find out to your own demonstrated satisfaction that the burden *weighs at least five hundred-weight*. You sling it off and give it a malicious kick, with the result that you break a thermometer or some such delicate instrument. Then you try to walk, but stagger about like a drunken man; there is no small to your back, your back tendons are puffy and tired like those of an old horse, your head swims, and your eye is dim. Patience and rest, however, gradually bring you round, and soon you regain strength and spirits in feeling that at least you have conquered a day's difficulties and have brought your board and lodging so far with you.

Ah! think of it, you knapsack mountaineers, you feather-bed Swiss mountaineers, with your tracks, your hotels, your guides, your porters, and your huts. No; this New Zealand work is not like yours.

But then, you see, we are enjoying what you cannot get. Exploring and opening out virgin fields, learning to be our own guides—and porters—from that best of masters—hard experience.

We struck up the little valley which here exists between the lateral moraine on our right and the hill on our left, and toiled on amidst dense scrub so gnarled and matted that we could at times walk on it as on a spring bed, though now and then going through, of course. The scrub alternated with slopes of loose strips of moraine. By evening we reached a little blue lake which feeds the creek issuing from the valley's mouth, and here we pitched our tent for the night.

The sub-Alpine vegetation here is interesting and varied. Wild Irishman (*te matakuru* of the natives or *matagourie* of the shepherds), Spaniards, with leaves like carving-knives and points like needles, having stalks sometimes eight or ten feet high; stunted totara, many varieties of veronica, celmisias with large marguerite daisy-like flowers, the beautiful white ranunculus, and a hundred bushes and creepers all mixed up in the most glorious confusion amid rocks sometimes covered with slippery moss, over and amongst which it is anything but pleasant to force one's way. The mountain sides are clothed almost up to the snow-line with beech, totara, ribbon-wood, veronica, and other trees, the rich foliage being beautifully varied; but not having sufficient time to cut bedding, we spent an uncomfortable night. The first evening is always the worst in camp. In the morning we continued our rough journey up the valley and our struggle with the 'worrying' swag.

Soon we discovered traces of fires and old camps, and we knew we were on the tracks of Green's and Von Lendenfeld's parties. An hour for dinner under a splendid waterfall, and more toiling onwards, till at last we were over the last boulder-face from the mountain on our left, with the Ball Glacier in full view. Fox, bending down, picked up a portion of an old veil, shortly after I found a goggle box, then came a tomahawk lying on a rock, then the historical tent poles of Mr. Green, and we knew we had reached 'Green's fifth camp.'

Off came the swags, and right glad we were to be done with them. If a man were only built on the same lines as a Mount Cook grasshopper he might 'stand some show' in those parts, for these insects are the most accomplished rock acrobats, jumping twenty or thirty times their own length at a spring, landing on their heads or anyhow with a bang, and squaring up for the next jump as coolly as cucumbers.

We found many relics of Green's and of Von Lendenfeld's parties, amongst them a surveyor's chain, which, with Green's tent poles, we have for the last five seasons used to pitch our tents.

Scarcely were we made snug for the night when down came a terrific nor' wester, blowing with fearful violence, making the tent boom and shake till we expected it to blow to ribbons. Rain poured down, thunder, lightning, and avalanches all lent their aid, and the elements seemed to be having a generally rowdy time of it. All this, of course, meant snow on the higher peaks; our spirits fell to zero very quickly, and we gave up all hope of tackling Aorangi for at least a day or two.

The nor'-wester is the *Föhn* wind of New Zealand, similar in character to the *Föhn* winds of Switzerland or the *Pampiero* of the Andes. Warm air laden with moisture travels from the equatorial and Australian waters, till, striking the range of the Southern Alps, precipitation ensues, the wind descending on to the eastern plains dry and hot.

Having studied Von Lendenfeld's map of the Tasman Glacier and its surrounding peaks made in 1883 we knew our whereabouts; but as yet we had not seen the peak of Mount Cook, having been toiling up close under the eastern flank of the range, which continues from the peak proper for a distance of ten or twelve miles in a south-easterly direction.

The morning broke beautifully clear, and we were

early aroused by some inquisitive keas, or mountain parrots, which perched on the tent and set up an unearthly screeching. These birds are ridiculously amusing and tame, and we frequently replenished our larder with them by the aid of a shanghai, or common schoolboy's catapult, with which instrument of warfare I have the rather questionable credit of being somewhat of an adept. When I think of the savoury fries and stews which the shanghai has brought to our camp table—the table being usually a rock or a large lily leaf—I begin to be reconciled to the haunting regrets for apple-destroying and window-smashing which so often beguiled the tedium of a scholastic career.

We determined not to attempt any climbing so soon after the storm, but set out to reconnoitre the route taken by Mr. Green.

Mounting the steep lateral moraine of the Ball Glacier we were soon across it and on to the clear ice of the Hochstetter stream beyond, and felt the joyful crunching of our well-nailed boots as we tramped along over the uneven surface.

There is something exhilarating in this setting foot on the clear ice after days of clambering over cruel rocks, something that seems to thrill one as the nails go 'crunch, crunch' and give such grand foothold, a cheerful ring in the clink of the ice-axes, a peculiar charm in the tinkle of the little surface streams, a sense of peace and loveliness in all around, an inspiration of awe and grandeur in the glorious masses of mountains which rear their hoary heads for thousands of feet above, whilst over all there seems to hang an invisible and imperious over-ruling and omnipotent Power directing the marvellous workings of Nature. Here man may feel his littleness and his unworthiness, and yet with Byron he feels what is so beautifully expressed in 'Childe Harold'—

> I live not in myself, but I become
> Portion of that around me; and to me
> High mountains are a feeling.

The Hochstetter Glacier is one of the most impressive and beautiful sights in the Southern Alps. Its supplies come even from the very summits of Aorangi and Mount Tasman, the two noblest mountains in Australasia. Avalanches from the eastern and northern slopes of Aorangi descend to a large ice plateau situate at an altitude of 8,000 feet. From between the great north-eastern spur of Aorangi and the southern slopes of Mount Tasman the Linda Glacier issues also into this plateau; it was discovered and named by Mr. Green. From the eastern slopes of Mount Tasman and the southern flanks of Mount Haast avalanches also descend to the plateau, which must be some ten or twelve square miles in area. This plateau has but one outlet—the fall of the Hochstetter Glacier. Viewed from below, the frozen cascade tumbles in the wildest confusion over a precipice of 4,000 feet to join the Tasman Glacier at an altitude of 4,000 feet (roughly speaking), and presents a most wonderful appearance. The fall at the top is probably about a mile and a half in width, narrowing to one mile at its foot, and the ice is broken up into *séracs*, cubes, pinnacles, and towers of all shapes and sizes, intersected by crevasses of the divinest bluish-green colour, and each pinnacle crested with a white cap of unconsolidated snow. One enormous rock protrudes through the ice in its southern and lower portion, crowned with toppling *séracs* 200 or 300 feet in height, which at regular intervals fall over the face of the rock and descend in magnificent avalanches. First comes a report like a pistol shot, then follows an almighty crash accompanied by clouds of snow and ice dust, succeeded by a low rumbling thunder as the blocks expend their impetus on the gentler slope below, and finally settle down again into solid ice, to continue their journey of centuries towards the terminal face of the glacier nine miles down the valley. Above the fall stand out, in bold relief against the clear sky, the giant forms of Aorangi and Tasman.

To stand before this wonderful piece of Nature's work and gaze on the weird and fascinating forms of the attendant peaks is an experience not to be forgotten.

The awful and solemn silence of the mountains, broken only now and again by the crash and thunder of an ice avalanche or the screech of a solitary kea, the complete desolation, the loneliness and remoteness from the haunts of men, all tend to inspire one with deep thoughts and feelings. One line in Walter C. Smith's 'Hilda' expresses more than pages of mine would do—

> The silence of the mountains spoke unutterable things.

In two hours' time we were across the glacier and on the point of the ridge descending from Mount Haast, which bounds the northern side of the ice-fall. We began the ascent of the ridge amongst snow-grass and lilies, but soon the vegetation gave way to rockwork, and when a height of about 5,000 feet was attained we made sure that this was our correct route, and, mist coming on, we descended again, and reached our Ball Glacier camp in the evening.

We resolved to make our attempt on the peak early the following morning, and accordingly, at 5 A.M. packed our swags, containing 'tucker' for three days, spirit lamp, blanket, opossum rug, mackintoshes, instruments, a change of warm clothing, &c., intending that night to find a bivouac at 8,000 feet if possible.

Starting at 5.20 A.M. we crossed the Ball Glacier in the very dim light of a waning moon, and were on the Hochstetter ice at peep of day, and making good time across, reached the point of the Haast spur in an hour and three-quarters. A thick mist hung over us, and we waited for an hour for it to lift, amusing ourselves by smoking and botanising, and watching the antics of some queer little wrens. These birds are absurd-looking little creatures with long legs and longer toes, plump buff-coloured breasts, no tails, staring little eyes, and look for all the world like boiled potatoes with their jackets on, set up on hairpins and let loose on the rocks.

As the mist cleared we tackled the ascent, and found it pretty stiff work, although we had snow-grass to assist us for some way up; but the rocks above this began to show signs of rottenness, and much care was required to avoid dislodging them. We made good progress to about 5,000 feet, when we were quite baffled for a time, and were forced to leave the main *arête* and look for a more promising route on our right. Here we proceeded cautiously, crawling through a narrow niche in some overhanging rocks with a precipice of some hundreds of feet below. Then the climbing improved till our view upwards was bounded by an indefinite saddle in the rocks, which might have led to anywhere, but which did lead, as we subsequently found out, to the easy snow slopes above.

As the day advanced small falls of stone occurred, which caused some annoyance and danger, but we managed to avoid being struck by any. Then followed another stretch of rotten rock which Fox absolutely declined to tackle, and as it could not be turned by a détour we were brought up on this route.

Fox suggested descending again to cross a large glacier coming down from the ridge on our right, and trying the rocks on its opposite side. This plan we eventually carried out, but it was a fatal mistake as far as climbing Aorangi was concerned. Descending for about 1,000 feet we stepped on to the ice of what we then thought was the lower part of the Linda Glacier—owing to a strange error in Von Lendenfeld's map—but which in reality was the Freshfield Glacier. We put on the rope and our goggles, both indispensable in crossing such a snow-covered ice stream.

On taking to the rocks on the other side we soon gained the lowest ice slopes, covered with six or eight inches of snow in splendid order, and adhering well to the ice; now and then we took to the rocks, but climbed mostly by the snow slopes till we reached the crest of the ridge and looked over a precipice to Mount Haidinger and the Haast Glacier below.

It was now 11 A.M.; and after a short rest, upon my suggesting a move upwards, Fox said that he did not fancy the rocks above—which certainly did look bad—and counselled a retreat. Of course I was disappointed, and reluctant to give up the attempt so soon, yet there did seem to be no end to the difficulties above, and experience has since taught me that Fox was wise in his counsel, for it was indeed simple madness for two greenhorns to tackle such work.

I soon forgot my troubles in gazing on the scene which burst upon us as we gained the ridge. Below lay the major part of the Haast Glacier, descending in a similar manner to the Hochstetter ice-fall from the corniced *arête* of Mount Haidinger, a marvellous mass of *sérac* ice. A long rest here, and a resolve to revisit the locality during the next season with a stronger party, and we began the descent.

My first experience of glissading on the snow slopes below was decidedly amusing; but the art is easily acquired, and after the inevitable spill or two one soon gets into the way of putting one's axe directly behind and not at the side, as is the first impulse. Many and many a good slide have I enjoyed during the last six years, and I know no more exhilarating sensation.

In winter time, on the front ranges, we have sometimes made glissades of 2,000 to 3,000 feet without a stop, and on one occasion, in crossing the Mount Cook Range, Mr. Arthur Harper and I glissaded close on to 4,000 feet with only occasional stoppages for crevasses.

MOUNT TASMAN (11,475 FEET) AND THE HOCHSTETTER ICE-FALL.

Wheeler & Son, Photo.

after which we returned to extricate the buggy, which had come to a standstill on its side, and was fast being silted up with moving shingle. It required all our strength to free it, and in doing so one of the wheels 'buckled.'

I have no doubt that we presented an amusing and half-drowned appearance as we stood on the bank and called the roll. All that was missing was my mackintosh, a mat, and whip.

Then we jumped on our buckled wheel till it sprang back into its normal shape, and splicing up the harness, wended our way back across the minor streams to the track at Birch Hill, wetter, sadder, and wiser men.

We reached Pukaki Ferry an hour after dark and Fairlie Creek the next evening, where we found the township in a state of jollification over the annual race-meeting. Most of the New Zealand country townships boast of their annual race-meeting, the racing lasting one day, and the whisky part of the proceedings generally running into three.

Then we took the train for Christchurch.

CHAPTER IV

SECOND ATTEMPT TO CLIMB AORANGI

A Flooded Camp in the Tasman Valley—Hard Struggles with Bad Fortune—We reach Green's Bivouac

> If at first you don't succeed,
> Try, try, try again.—*Nursery Rhyme.*

During the winter following my first essay at Alpine climbing I was not idle, but made several pig-hunting excursions amongst the foot-hills in North Canterbury, in addition to which, with a companion in the shape of an old friend and schoolfellow, Mr. M. J. Dixon, I made the ascent of Mounts Torlesse (6,434 feet) and Puketeraki (5,780 feet) at a time when these mountains were snow-covered to within 2,000 feet of their respective bases.

The former ascent was accomplished in the face of a nor'-west gale, and well I remember how we had sometimes to lie down on the snow and hold on to our sticks to avoid being blown clean away. We have twice since climbed this peak under similar conditions, and I never remember the wind blowing with such force as it does on Mount Torlesse.

It was on February 1, 1887, that Messrs. M. J. Dixon, C. H. Inglis, and myself left Christchurch for a second try at Aorangi.

We were now well equipped for the attack, having obtained 160 feet of Alpine rope, three good ice-axes from M. Fritz Boss of Grindelwald, and suitable nails for our boots. Inglis had his camera and two dozen plates.

On arriving at the Hermitage we found that the Hooker River was up and quite impassable for horses, consequently we were forced to cross the Mueller Glacier by the Hermitage, walk up the Hooker Valley, and cross the terminal face of that glacier on to the western slopes of the Mount Cook Range, after which we worked our way down the river till opposite the Hermitage again, where a length of fencing wire was thrown across the torrent by which we were able to take our swags over.

The roar of the torrent was deafening, and oral communication across was quite impossible. The wire on our side was made fast eight or ten feet above the water, and on the other about twenty feet. Three cheers were given us by the party of tourists on the other bank, to which we replied, and then we were cut off from the haunts of men for a week, and thrown quite on our own resources for clothing, food, and shelter—board and lodging, in fact.

Then came the arranging of swags, adjustment of carriers, &c., and we soon discovered that we had all we could carry—over 50 lbs. each. Then followed the toiling down the steep bank of the river to reach the end of the range, in the piping heat and glaring sun, now and then having to ascend the slopes to avoid the river, which rushed along close to the rocks.

At one place in particular we experienced some

difficulty, having to resort to the use of the rope to climb a ditch or *couloir* in the rock-face where the river boiled past at a terrific pace. Here the camera was accidentally dropped, and falling down fifty feet or so, lodged on a ledge which overhung the water. Strange to say, when recovered it was found to be quite uninjured!

By dint of continued exertion and considerable expenditure of adipose tissue we at last turned the end of the range, and upon reaching the first water as we struck up the Tasman Valley, boiled the 'billy' and made a good lunch.

The wind now began to rise from the nor'-west, and clouds of dust were sweeping down the valley, so we lost no time in pressing on to a patch of Irishman scrub a mile or so below the terminal face of the glacier. We hurriedly cut some bedding and pitched the tent before the rain came on, in rather close proximity to an old creek-bed, which had apparently been dry for some time.

That creek made up for lost time during the night, and soon the rain came down in bucketsful as we lay our wearied limbs to rest in our oiled calico blanket-bags. The thunder crashed and the lightning flashed, and the Tasman River began to roar, and by one o'clock such a quantity of rain had fallen as to convert the dry creek-bed into a roaring torrent, whose waters threw up a bank of shingle, and, turning its course (horror of all horrors!), came right into our tent. In less than a minute from the time that we felt the first trickle there was a foot of water in the tent, and all our impedimenta of every description were sopping or floating about in the dark, and in imminent danger of being washed away.

Hurriedly we collected all we could into our blanket-bags, got into our boots somehow, and made for higher ground. We could not see a rise in the ground, but after wading about found a small portion out of water, and, with much strong language and trouble, succeeded in repitching the tent—after a fashion.

Ah! well do we remember the miseries and discomforts of the scene. Wind blowing in fitful gusts, rain coming down in sheets, while thunder and lightning and the incessant roar of the Tasman all tended to make the scene one of terror and discomfort. Matches nearly all destroyed; bread reduced to a state of pulp; blankets and clothes wet; instruments, boots, ropes, ice-axes muddled up anywhere, some in the tent, some being silted up or washed away from the spot where the tent was first pitched; the floor of the tent now hard, wet stones, in lieu of comfortable, dry tussock. Oh, the misery of it!

We lay in our wet clothes the rest of that night, all the following day, and the next night. Inglis and I scarcely stirred but to eat some disgusting, soppy mixture or to light our pipes; but Dixon pluckily rigged up a breakwind with an old tent left by the Birch Hill shepherds, and after three hours' persistent labour kindled a fire, improvising a chimney out of a pair of white flannel trousers and sundry other garments!

We were quite hemmed in by water, and were in a constant state of anxiety lest the river should make

depredations in our direction, as it was quite close to
us, whilst in the creek on the other side we could hear
the rocks being rolled down by the force of water.

Nine inches of rain had fallen during the forty-
eight hours, but on the Sunday it cleared, and once
again the warm sun shone out, the clouds drifted away
from the mountains, the birds began to sing, and the
waters subsided as quickly as they had risen, and our
spirits rose again as we spread out our wet belongings
on the scrub and donned a shirt, hat, and a pair of
boots apiece, and set out for a visit to the scene of
devastation at the face of the glacier whence the river
issues. The costume was airy but convenient, as we
had to cross several streams before reaching our destination.

We were well rewarded for our walk, for a wonderful sight was presented where the river flows out from
the glacier. For a distance of half a mile from the face
the banks of the main stream were strewn with blocks of
ice of all sizes up to twelve or fifteen feet in thickness.
At one spot the river rushed in mad violence from a
great cavern of ice; in another it rose as from a geyser
from under the ice, sending up a large column of water
to a height of six or eight feet.

It was quite a new sensation to be dry again, but
that night rheumatism screwed my joints, and some
venomous insect bit my shoulder, causing intense pain
for a short time.

While the rain continued we had all thought of
falling back on the Hermitage as soon as we were
able, but a bright sunny morning caused us to change
our plans and forge ahead for the Ball Glacier

camp, weakened though we were in strength and supplies.

Already we felt that our chance of ascending Aorangi was gone, for the snow lay thick on the upper peaks and avalanches were of common occurrence; yet we doggedly pushed on, determined not to turn without a struggle.

Leaping from rock to rock, avoiding the scrub and Spaniards by sticking to the moraine slopes, and scrambling over great *tali* of boulders which came from the mountain sides, by evening we reached our destination (the Ball Glacier), and finding the surveyor's chain, tent poles, and hatchet—left by Fox and myself the previous season—in good order, we quickly had a comfortable camp pitched. A small army of mountain parrots or keas soon assembled, and the unerring shanghai procured grilled kea for supper.

Next morning broke gloriously fine, and by 7 A.M. we were away with blanket-bags, three days' 'tucker,' and a change of warm clothing, intending to reach Green's bivouac on the Haast Ridge that evening, and to make a final dash at Aorangi on the day following.

Once again we plunged into all those pleasures and joys of mountaineering. Again we felt the clear ice of the beautiful Hochstetter Glacier crunch under our iron-shod feet. Now we were away from all the hum-drum cares of life, from the misery of flooded camps, in the free mountain air, with the stupendous ice-falls and the majestic peaks all around. We seemed to breathe a heavenly atmosphere, to live a new life in another and a better world. Where is the man who can come into contact with these surroundings and not be better in body and soul?

We reached the foot of the Haast Ridge by 9.30, and here we debated as to whether we should tackle Aorangi after all, or try De la Bêche, further up the glacier (which peak would be an easier ascent and command a magnificent view of both eastern and western glacier systems). Aorangi it was, however, we had come to tackle, and so, again shouldering our swags, we went at the ridge.

We kept to the crest of the spur and found the climbing very simple, for a thousand feet amongst lilies and snow-grass; but after that the real business amongst rotten and precipitous rock ridges and faces commenced, and we had to put on the rope. At this time none of us were very proficient in the use of the rope, but we soon began to value the assistance it affords and to appreciate the assurance it inspires.

It was not until 5 P.M. that we reached the top of the ridge, where we soon discovered Green's bivouac, not far from which spot we determined to spend the night.

All the way up we had been climbing with the Hochstetter ice-fall on our left, and had been favoured with the grandest views of Aorangi, which looked absolutely impregnable; but as our view of the Linda Glacier and the Great Plateau was shut off by the upper part of the Haast Ridge, we could not see the route which we were bent on following.

Here I may remark that the route by which Mr. Green, and subsequently Dixon and myself climbed the mountain cannot be seen from any distant point. I refer, of course, to the upper part of the route above the Haast Ridge. Even the plateau is so shut in as to be invisible from any distant point, except from the peaks of the Malte Brun Range on the opposite side of the valley.

Scraping away all the larger stones from under an overhanging rock and building a semicircular breakwind, we dug holes for our hips (one gets very sore in hard beds of this nature if such a precaution be neglected), wriggled into our blanket-bags, boiled a pannikin of Liebig, and slept like tops till the morning.

The rosy fingers of the morn had just opened the gates of day as our heads emerged from the apertures of our bags, and showed one of the most magnificent panoramas of Alpine wonder which it has been my lot to view.

Three thousand feet below us lay the Tasman Glacier with its marvellous stream of pure ice, on our right the Hochstetter ice-fall, on which we could look down and view with wonder its chaos of *séracs* and crevasses, the ice-clad precipices of Aorangi rising heavenwards from it in bold ruggedness. Down the valley to the southwest the grey moraine, with the meandering river still further afield. Across the valley the rocky peaks of the Liebig and Malte Brun Ranges with their hanging glaciers, and right opposite to us Malte Brun himself, a pyramid of red rock, flanked by ice and snow slopes, standing out clearly against the morning sky like a great grim castle, and looking quite safe from any assault of man—on this side at all events. Following round the panorama to the northwards, Mount Darwin sends its one great glacier sweeping down into the main stream; then the Hochstetter Dome stands at the head of the Tasman Glacier itself, and westward rise the noble summits of Mounts Elie de Beaumont, Green, and De la Bêche—the last a most beautiful triple peak, queen of the whole group, and over 10,000 feet in height. Still following round, the eye falls on the Rudolf Glacier descending from the peak of the same name, then Mounts Jervois, Spencer, Glacier Peak, and lastly Mount Haidinger, a fine flat-topped mountain clothed from base to summit in broken ice.

Behind us lay Mount Tasman (11,475 feet), invisible over the higher parts of the spur on which we were now situated. From our coign of vantage we counted twenty-five tributary glaciers of the Tasman, some with ice-falls, others joining with graceful curve.

We congratulated ourselves that all our weary toil and hard swagging had not been fruitless, and felt quite compensated for the miseries we had gone through at the lower camp, though the main object of our visit, we feared, was about to be defeated in a very short time. We pulled ourselves together, put on the rope, and resolved to make some pretence of a fight for it.

After an hour's work we reached the highest rocks, then there came a dip on to a snow saddle, beyond which, again, snow slopes lead on to the final summit of the spur which hid the Great Plateau.

But it was not to be; for whenever we went on to snow we sank waist-deep, and struggled in vain to make any headway. Here, then, we were beaten, and planting our Christ's College flag in the highest rocks, gave it three cheers for the old school days, and depositing a bottle with the record of our ascent, turned our backs on the grim giant Aorangi, and began to go down.

We struck a better route down by going into some *couloirs* north of the *arête* of the spur, and reached the Ball Glacier camp again, going down the following day to the Hermitage, after crossing the Hooker by the kind assistance of a shepherd from Birch Hill. The Hooker River had risen to such an extent during the rain storm as to carry away the wire on which we had slung our swags across. The camera was warped with the wet at the lower camp, whilst the plates were anything but 'dry' after the storm, so photography was altogether a failure in this excursion.

In the winter time we amused ourselves with another ascent of Mount Torlesse.

CHAPTER V

THIRD ATTEMPT TO CLIMB AORANGI

Photography on the Tasman Glacier—Attempt to scale Mount De la Bêche

Where rose the mountains, there to him were friends.—*Childe Harold.*

IT is a thousand pities that the ennobling pursuit of mountaineering is so neglected in this wonderland of peaks and glaciers. Such advantages as we enjoy surely cannot exist much longer without calling out the spirit which lies dormant in hundreds of the lovers of adventure and worshippers of the beautiful in Nature, who live on in our midst from day to day in a conventional and monotonous round.

There are pleasures in the pursuit of adventure amongst the great snow-fields and glaciers which only those who are initiated can thoroughly enjoy.

Ask the man who goes climbing what these pleasures are, and he cannot tell you, he cannot define them—yet he feels them, and they are ever luring him on. They are indefinite, inexpressible; but there is a sort of 'mountain fever' which comes when one has once 'lost one's heart to the great mountains.' In the work all a man's best physical, and many of his mental, powers are brought out and strengthened. There is the energy, perseverance, and patience to last through a long day's swagging, the pluck to face all sorts of dangers amongst the snow, ice, and rocks, combined with the prudence to know when, for the safety of oneself and the party, to give in and restrain enthusiasm. There are the qualities of organisation and system, for which plenty of exercise is found; indeed, one cannot overrate the benefits which accrue.

Let any who have indulged in different branches of athletics put their swags on their backs and go for a mountain climb, and I venture to say that there are greater opportunities for bringing their frames into good going order and testing their muscular abilities than can be met with in any school of athletics.

I have known men in England who have revelled in all our great national games, but who invariably put mountaineering at the head of the list after once having tasted the sweets of climbing and been captivated by the charms of the world above the snow-line.

To the artistic what do not the mountains offer? To the botanist, the geologist, the naturalist, the athlete, and even to the invalid? The strange new world one enters in sub-Alpine regions, the 'foretaste of heaven' one seems to get above the snow-line.

In out-of-the-way New Zealand we have all these benefits at hand, and yet we leave the opening out and exploration of our great glacier systems to foreigners and to visitors from distant lands.

But this is digressive, and I must tell the story of our third visit to the Tasman Glacier.

On the evening of March 23, 1889, the visitors at the Hermitage were suddenly moved to compassion, mingled with no small amount of amusement, in beholding through the fast-falling snow-flakes the arrival of a dog-cart and tandem.

The leader of the team, a big chestnut draught-mare, seemed to be doing all the work, and pulling along wheeler, cart and all. The travel-worn and weary occupants of the vehicle were Mr. M. J. Dixon and myself, and we had taken French leave for Mr. Huddleston's chestnut at Birch Hill, six miles down the road from the Hermitage, our leader having almost given in after a 250-mile journey from Christchurch.

Another bold, would-be mountaineer, Mr. P. H. Johnson, accompanied us with the knocked-up leader, and following in the coach was Mr. F. Cooper, a photographic operator from Messrs. Wheeler and Son of Christchurch, who was to join our party for a week's work amongst the scenes of the Tasman Glacier.

The morning of the 24th revealed the flats around the Hermitage all snow-covered, and the day was devoted to completing preparations for a fortnight's camp on the glacier.

On the 25th, the weather improving, our party left the Hermitage, being joined by James Annan and William Low, the former a boundary keeper on the rabbit fence, the latter engaged to help us with the swagging. Two better men over rough ground never put swag on back, and both entered into the spirit of the expedition and worked like Trojans to make it a success.

We drove our dog-cart down to the Hooker River at the usual crossing-place—the point of the Mount Cook Range—over two or three miles of boulders which tested the merits of the coachbuilder's art to the utmost, as also the driver's ability to stay in the cart. Here we found that a wire rope, some 200 feet in length, had been thrown across the river to facilitate the work of the rabbiters, who were engaged in keeping back the hordes of 'silver-greys' which were making their way northwards and ruining run-holders right and left. On this wire rope is slung, on runners, a rude box, travellers entering the same pull themselves across, and almost invariably take the skin off their knuckles with the runners. Crossing by this rope we piled our swags on to Annan's packhorse and walked three miles up the valley to a patch of Wild Irishman scrub, where since our last visit a small galvanised iron hut had been built. A day's delay here with bad weather, and then we shouldered our swags, and on the evening of the 27th reached our well-known Ball Glacier camp.

Our plans were as follows: To do a few days' work with the photographer, so as to settle his business first, and then be free to tackle Aorangi during the following week. We wished to give the photographer every assistance in our power, as such scenery does not often come within reach of the photographic artist, however energetic he may be, and can only be approached by a properly equipped Alpine party, strong enough to carry a good supply of provisions and all the necessaries for preserving life in such out-of-the-way parts.

Our first excursion, then, was to cross the Tasman Glacier and make for the point of the Malte Brun Range at the turn in the glacier just opposite the point of De la Bêche. Here it was that Dr. von Lendenfeld had made his bivouac for his remarkable ascent of the Hochstetter Dome in 1883, when he was accompanied by his wife and one porter—an ascent that took twenty-seven hours of constant ice and snow work. This excursion would effect the double purpose of giving us some practice in ice work, and of securing a fine set of views.

The day was gloriously fine, and we felt our spirits rise as we scrambled over the massive lateral moraine of the Ball Glacier, across the glacier itself—which, by-the-by, shows very dirty ice at this point, being laden with rocks brought down many years since in the avalanches from the great ice-seamed crags of Aorangi, which towered in lofty grandeur above us—then over the medial moraine between the Ball and Hochstetter Glaciers, where a halt was made, and views of Aorangi and the Hochstetter ice-fall were secured.

Once more we stood before this marvellous piece of Nature's handiwork, again we heard the thunder of the avalanches, again we saw the glinting, bristling *séracs*, and gazed in silence and admiration on the ice-fall of the Hochstetter.

Crossing the Hochstetter we struck up the medial moraine between that and the Tasman, straight for the point of De la Bêche.

The best walking on the New Zealand glaciers is almost invariably found upon the margin of the medial moraine close to where it joins the clear ice, so that one is travelling over a mixture of ice and rocks. The clear ice is too hummocky and entails much undulating progression, if I may use such an expression, and

Hooker Saddle

[Wheeler & Son, Photo.

MOUNT COOK AND THE HERMITAGE

the moraine itself—well, the walking on the moraine itself cannot be fitly described in parliamentary language.

We secured many good views as we proceeded with a 10 × 8 camera. Mount Haidinger on our left was particularly fine, its eastern face being almost entirely clothed with the Haast Glacier, which struck us as being one of the finest cascades of ice we had yet seen, larger in extent than the ice-fall of the Hochstetter, though not so picturesque.

Time was fast going, and we found that to get off the glacier before dark it would be requisite to strike away to our right, over a mile of much crevassed ice, to the gully next in the Malte Brun Range, which we had originally set out to reach. Jumping crevasses and cutting a few occasional steps, we at last arrived at the eastern side, finding a very suitable place to pitch our Whymper tent, and discovering to our joy a small supply of firewood.

The gully in which we camped had its origin far away up in the red-sandstone precipices of Malte Brun, and in its bed rushed down a foaming mountain torrent fed everlastingly by the many small hanging glaciers above. This stream rushed headlong into a large tunnel of ice in the side of the Tasman Glacier, over which was formed a tremendous cave, above which, again, were sheer walls of ice capped with morainic accumulations, the height from tunnel mouth to moraine summits being about 500 feet.

A view of this cave was secured by the photographer.

Friday the 29th was a morning to be remembered.

Thick mists covered the peaks and seemed to hang over us like a pall. Here and there a shaft of sunlight penetrated to the ice-field at our feet. Only now and then would the rude screech of a kea remind us that we were not really dreaming in some enchanted land.

We had often talked of attempting the ascent of Mount De la Bêche when we should have polished off Aorangi; but as Aorangi seemed to require so much 'polishing off,' and we were now camped so close to De la Bêche, we thought we might as well try our hand at the mountain and see what we could do in a one-day's trip from this point, while we left the artist to his own devices for the time being.

De la Bêche, then, it was to be. So off we started after a breakfast of sheep's tongues and Liebig, putting our oilskins on our backs and taking our axes, and striking due north for the foot of the long *arête* which descends from the mountain and separates the Rudolf from the Tasman Glacier. Halfway to our ridge we had to put on the rope, for legs began to go through the now snow-covered crevasses in a promiscuous and unpleasant fashion.

It was indeed like an enchanted land, for the atmospheric effects were extraordinary. High up, shadowed in the mist, were reproduced the forms of the highest peaks of Mounts Malte Brun and Darwin. There was no mistaking their familiar outline, which was thrown out in the mist thousands of feet above, like the spectre on the Brocken.

Then the atmospheric effect of the mist hanging over the Rudolf Glacier was most wonderful. Looking

CROSSING THE HOOKER RIVER.

[Wheeler & Son, Photo.

up the glacier, we seemed to gaze into an enormous blue grotto, the sides being the slopes of the main chain with all its broken glaciers, and the western slopes of De la Bêche, whilst the overhanging mist furnished the roof or ceiling. A soft, warm, blue colour pervaded the whole, beautiful beyond expression.

Arriving at the foot of our mountain we commenced the ascent, finding the snow of the ice slopes in a loose and powdery condition, and having to exercise much judgment to avoid precipitating avalanches in the steeper pinches.

We climbed without the rope, rapidly, and alternately in snow and rocks, finding the latter very good —mostly of a red sandstone on which the nails of our boots took good hold. Looking now and then at the aneroid, we began to feel confident of making the ascent and returning to our camp by nightfall. But it was not to be, for, at an altitude of 8,100 feet, we were brought up by a very bad *bergschrund* and ridge of rocks succeeding it.

To the unlearned in Alpine parlance perhaps an explanation of the nature of a *bergschrund* is necessary. At the upper termination of nearly all highly situated ice slopes there almost invariably occurs between the rocks above, or between the ice slope and the permanent clinging ice above, a large gap or crevasse, partially filled or bridged with new snow during the winter months, but more open as the warmth of spring and summer causes the snow to melt and the ice to shrink away.

This crevasse or gap is called a *bergschrund*, and occasionally one may find in it places where the ice nearly or quite touches the rocks or ice of the upper side, or sometimes a sound snow bridge may be discovered. These bridges afford the only means of crossing wide *bergschrunds*. At the place in question a sharp ridge of ice, the lower lip of the *bergschrund*, led on to a frail snow bridge with a dip of some six feet or so in the centre, over a bottomless abyss some fifteen feet wide.

Dixon cut steps along the ice ridge, having first to remove a foot of fresh snow from the surface, and then we walked this novel tight rope, the *bergschrund* on our left and steep ice slopes on our right, and crossed the bridge in safety to a small ledge of ice where there was only just room for three to stand. Could we proceed? The rocks above were very bad and ice-coated. I went at them, clearing the inch or so of ice to get my fingers into chinks in the rock, and 'squirming' up on my stomach, clinging with toes and fingers, and feeling disposed to hang on by my teeth or even by the proverbial eyelids, reached, fifty feet above, the crest of the ridge.

I had been in some queer places in the mountains, but, pardon the use of a colonial expression, this one decidedly 'took the cake,' and I shall never forget the start I received when I found myself looking over a sheer upright face of rock on to an unnamed tributary glacier of the Rudolf, 1,000, perhaps 2,000, feet below. I dared not stand up and could scarcely crawl, but lay full length on the steep eastern slope looking over the sharp ridge down the western precipice. On the right, the razor-like *arête* of rock continued upwards, and seemed almost, if not quite, inaccessible.

Then there was a long-range discussion between

Lowest Peak Middle Peak Eastern Spur
John Pass Glacier

AORANGI FROM THE BALL GLACIER

Dixon and Johnson on the ledge below and myself on the ridge, ending in a decision to descend.

I never to this day can imagine how I came down that fifty feet of rocks without slipping into the crevasse below, but, by the aid of Dixon's directions, I managed to find chinks in the rock-face for the toes of my boots, and reached the ledge to breathe the air of relief once more.

Here we held a council of war. We might, by a traverse of the ice ridge below, gain the rocks again above this bad place; but the summit was yet 2,000 feet above us, the cold so intense that the steel of one's axe would adhere to the hand, the time was fast going, and the photographer and our men would be much concerned if we stayed out another night, besides which we were short of provisions, our original intention having been to stay out but one night. We decided to acknowledge ourselves beaten for the time being and to return to camp.

It goes against the grain with Dixon and me to turn back beaten from a peak. Indeed De la Bêche and Aorangi are the only ones to which we have lowered the colours of our grand old school—Christ's College Grammar School, of Christchurch, New Zealand—and the latter we have since revenged ourselves upon. The former will not run away, and we are nursing a vindictive feeling against this noble triple-topped summit.

Descending very rapidly, glissading now and then in safe places, we reached the foot and struck over the Tasman Glacier again for our camp on the Malte Brun.

Well for us that we had turned from De la Bêche,

for an hour from camp, Dixon, who had been complaining of not feeling up to the mark for some days and had been lagging—an unusual thing for him—was suddenly seized with violent cramp in the stomach and thighs. We thought at the time it was only temporary, consequent upon great physical exertion and drinking too much snow-water; but unfortunately he did not seem able to shake it off, and we had some difficulty in reaching camp over the maze of crevasses which occur in the glacier just where our Malte Brun Creek enters.

Here was a nice state of affairs. One of our best men gone wrong. How about Aorangi next week?

Saturday morning found us 'tuckerless' and hungry, and Dixon worse rather than better.

At 9 A.M. we struck camp and started for the Ball Glacier—really only four hours distant. Whilst taking some views an hour from camp we suddenly heard shouts down the glacier, and found that it was our trusty men, Annan and Low, who, being concerned about our lengthened absence from the lower camp, had come out to look for us.

Johnson, Low, and Annan took the bulk of the swags and started independently for the Ball Glacier, whilst I stayed to follow at a more leisurely pace with Dixon and the photographer. Dixon could only walk for a few minutes at a time and required to rest very frequently, so I sent Cooper on alone, not dreaming for a moment that he could go wrong in such simple ground, where no crevasses to speak of occurred.

It was 5 P.M. ere we arrived at the head-quarters after a gallant struggle on Dixon's part. These are the times which test a man's capabilities, these are the

Aorangi Tasman Haast Haidinger

ICE CAVE, TASMAN GLACIER

[*Wheeler & Son, Photo.*

trials of endurance to which the unfortunate who chances to be taken ill in these Alpine regions is subjected, and it was a great relief to all to see the afflicted one struggle bravely into camp.

But a new trouble arose. There was no photographer, and he ought to have turned up long ago. Johnson set out to look for him, and after an absence of an hour I was just putting up a swag of mackintoshes, provisions, &c., prepared to spend the night photographer-hunting on the glacier, when Johnson's figure appeared against the sky on the crest of the lateral moraine, shortly followed by that of the missing man, who had wandered down past the camp instead of turning off at the right place. Low and Annan had gone down the valley, and were to come up next day with more provisions.

The next day being Sunday, we decided to have a day's well-earned rest. Messrs. Brodrick and Sladden, of the Survey Department, came up with Annan and Low to dinner, bearing with them medical comforts for the use of our invalid.

As there were still some dry plates left unexposed, Cooper and I went out about 10 A.M. and climbed to a height of 1,000 feet above the camp, on the Ball Glacier spur, from whence we secured a panoramic view on four plates of the glacier and the mountains opposite.

From this point, seeing Aorangi looking so grand, we pushed on up the ridge, intending to secure an exposure from a high altitude. Upwards we climbed, and the further we went the more I was lured on towards the main southern ridge of the mountain. I even conceived the idea of making a pass over to the Hermitage *via* the Hooker Glacier. But the work became more difficult, and we got into patches of snow and were unfortunately without our ice-axes. This made our progress more slow and cautious. Still we pushed forward, the scene becoming grander at every step.

At length the light began to fade, and I saw that to get an exposure of the peak from the main ridge was hopeless, so Cooper unlimbered his instrument and I pushed on alone, determined to reach the saddle, at least, and see over to the other side. Reaching the final snow—that covering the actual head of the Ball Glacier, which had been below us on our right all the day—I sped across it as fast as I could go, and keeping a sharp look-out for indentations indicating covered crevasses, reached the rocks of a peak situate a little south of the saddle of the Ball Glacier. Crawling over a snow bridge spanning the *bergschrund*, which crumbled uncomfortably under me as I laid hold of the rocks on the upper side, after a short scramble I attained the summit.

How shall I tell of the view southwards which met my astonished gaze? How describe the glorious sunset effects? Life is not long enough to attempt it.

I was on the nameless peak south of the Ball Glacier saddle at an altitude of 7,540 feet—the highest peak south of the great majestic mass of Aorangi himself, who towered up for another 5,000 feet above me.

I quote from Mr. Green to give some idea of what he thought of our mountains from this point:—

'Deep down below us lay the Hooker Glacier, re-

Rudolf Glacier   Mount De la Bêche   Mount Green   Mount Elie de Beaumont

MOUNT DE LA BÊCHE (10,021 FEET) FROM THE TASMAN GLACIER

[*Wheeler & Son, Photo.*

minding us of the downward view from the *arête* of the Finsteraarhorn, while beyond, the glacier-seamed crags of Mount Sefton towered skywards.

'Further off lay the *mer de glace* of the Mueller Glacier, a splendid field of white ice, its lower moraine-covered termination lost in the blue depths of the valley at our feet. The high ridge connecting Mount Sefton with Mount Stokes alone prevented us from seeing the western sea. It was a glorious day, scarcely a breath of air stirring; no cloud visible in the whole vault of blue; ranges upon ranges of peaks in all directions and of every form, from the iced-capped dome to the splintered *aiguille*. It was a wonderful sight, those lovely peaks standing up out of the purple haze; and then to think that not one had been climbed! Here was work, not for a short holiday ramble merely, not to be accomplished even in a life-time, but work for a whole company of climbers, which would occupy them for half a century of summers, and still there would remain many a new route to be tried. Here, then, we stood upon the shoulder of the monarch of the whole mountain world around us, within less than 5,000 feet of his icy crown, but a long, jagged, ice-seamed ridge lay in our path. Was it accessible? Let us see!'

It was not accessible, as anyone who has read Mr. Green's interesting book will know, and I could see from my standpoint very plainly that Mr. Green, with Emil Boss and Ulrich Kaufmann—two of the finest mountaineers in the world—could not do otherwise than accept a defeat.

Just such a scene as Mr. Green describes I saw,

only that its mystic beauty was intensified by the soft glow of evening as the sun sank lower and lower, at last dipping behind a bank of crimson clouds hanging over a saddle to the westward.

I seemed spellbound and almost riveted to the spot, and could only tear myself away when I realised the awkward position of the photographer and myself, trapped, as it were, by the fast-closing darkness, 4,000 feet above our camp, with all sorts of climbing difficulties below. Clambering down the rocks and jumping the *bergschrund*, away I sped over the *névé* slopes, and reaching Cooper after an hour's absence, found him just packing up his camera.

It is too long a story to tell of all our troubles and adventures in getting down the mountain in the dark; letting ourselves down on to the rocks, scraping our hands on sharp edges, plunging knee-deep in soft snow, following false ridges terminating in precipices down to the Ball Glacier below, retracing our erring steps, and at last coming to vegetation again; then going down off the ridge towards the Tasman, trying to hit the head of a long shingle slip I was acquainted with, hearing 2,000 feet above the camp the first 'cooee' from our anxious mates below, and getting down eventually at half-past ten, ravenous, and almost torn to pieces by the sharp rocks, Spaniards, and scrub.

Johnson—always self-denying and considerate for others—was out photographer-hunting again, having gone on to the Ball Glacier and shouted himself hoarse; he arrived back in camp at 1 A.M. (having been guided home by a fire which I had kept going on the moraine since our return), after having experienced a fruitless hunt of eight hours over rough rocks and ice. This finished the photography, and on the following day Cooper and Low went down to the Hermitage. A finer week for securing negatives could not have been wished for, and the thirty exposures resulted in the best set of mountain views yet obtained in New Zealand.

Now ensued a few days' rest, Dixon, Johnson, and I being left in camp with a week's provisions and designs on Aorangi, when Dixon should have recovered his strength.

Only one short excursion did Johnson and I make, to see if it were possible to reach the Great Plateau from the eastern buttress of the mountain, and so save crossing the Hochstetter Glacier and climbing the Haast Ridge beyond. Our endeavours were fruitless, for at a height of some 6,300 feet we were brought up by a high wall of rock. I still think, nevertheless, that the plateau could be reached in this manner when a good deal of snow fills the rocky *couloirs* or ditches which in places descend in this wall of rock. Should this be so, it will no doubt prove to be the route of the future for reaching the Linda Glacier and Aorangi.

The rock-climbing here, however, is very dangerous, as the frost has split the rocks up in all directions. One small stone thrown down from above sufficed to start many tons of loose matter in the *couloirs*, which rattled down to the glacier below, sending up clouds of dust in its descent.

CHAPTER VI

THE ASCENT OF THE HOCHSTETTER DOME

*Camp under De la Béche—Twelve Hours on Snow and Ice—
The Pangs of Hunger*

THURSDAY, April 4, was a memorable day, for Annan coming up from the Hermitage with a further supply of the ever-welcome 'tucker,' we started on one of the finest mountain expeditions I have seen in our New Zealand mountains.

It was not part of our original plan to ascend the Dome; we merely intended to reach the Lendenfeld Saddle and get a glimpse of the opposite coast and the western ocean, and it was with this object in view that Johnson, Annan, and I shouldered our swags and tramped off to the foot of De la Béche, which was made in three hours' hard walking.

Here we camped in a snug hollow between the lateral moraines of the Tasman and Rudolf Glaciers. Small shingle composed our bed, and a snow patch close by provided us with water, which we boiled in our 'Aurora' stove, as no firewood was to be found so far up the glacier.

A fine Friday morning found us at a quarter to seven on the rope, and making hard work of it amongst the crevasses of the Tasman Glacier.

PEAKS ON MALTE BRUN.

[Wheeler & Son, Photo.

I remember well how we resorted to all sorts of dodges to get over the difficulties, taking the snow slopes of the mountain sides here, cutting a few steps there, even going to the length of climbing down into crevasses and crawling under ice blocks. But eventually we passed the worst of the crevasses, and made good time over the smooth, snow-covered surface of the glacier.

The distance from our De la Bêche camp to the saddle must be about six or seven miles, but in the soft and treacherous snow it seemed more like sixty or seventy.

The glare was something dreadful, and soon our faces and hands were of the peculiar chocolate colour which invariably comes under such circumstances. We could not bear the goggles off for an instant. Gradually we rose as we plodded away, now and then stepping over an open crevasse or making a détour to find snow bridges. There are but few crevasses, however, for several miles, only when in the proximity of the saddle where the gradient increases they once more begin to occur.

On either hand fresh beauties opened out; De la Bêche on our left presenting the most wonderful face of *sérac* ice, streaked here and there with avalanche slopes, whilst on the right Mount Malte Brun—the Matterhorn of New Zealand—reared his great red precipices heavenwards, and further on the Darwin Glacier and Mount Darwin showed in a glorious light their magic splendour.

Now on our left we passed Mount Green, a fine precipitous cone of rocks and ice, and then we rose faster and faster as we edged on to the slopes of the great Hochstetter Dome on our right, whilst opposite, Mount Elie de Beaumont showered down his ice streams to join the Tasman.

Taking turns at leading, at last we came to what looked like the final rise. An exclamation broke from Johnson as he espied the new moon appear over the saddle ahead. It was a small matter, but it seemed to revive our failing energy and to call us on to victory to see the silver crescent apparently awaiting us on the snow ridge. Then a distant peak appeared—a wild cheer broke from us; another peak, and yet one more, followed by groups of twos and threes, dozens, hundreds—glaciers! forest! a river! the sea! the boundless ocean! 'Hurrah!' we shouted, 'our tramp has not been in vain.'

Here we were in the heart of Nature's solitudes, where only once before the foot of man had trodden the eternal snows.

We spent forty-five minutes refreshing the inner man and drinking in the glorious view, consulting maps, and reading the aneroid. The saddle was 8,600 feet high; the Dome was but 9,315 feet. Should we try it? Yes, we would.

At it we went, cutting many steps and crossing several awkward *bergschrunds*, until we reached a level plateau. Crossing this field we attacked the final slopes. It was terrific work, and the last pinch required 280 steps, all cut with the spike of the axe and deeply graven, as a slip in such a place would probably have meant the loss of the entire party in one of the crevasses in the slope below.

My hands were blistered with the axe work, but at 3 P.M. we were able to walk on the fast rounding-off slopes without steps, and soon we were on the summit, happy and flushed with victory. The mountain has a double top and we were on the western and slightly lower one.

What shall I say of the view from the Hochstetter Dome? It is comprehensive and wonderful. The whole country lay like a map before us. Westwards Elie de Beaumont and the western ocean, at our feet the Whymper Glacier, from which flowed the Wataroa River, threading its way through forest- and glacier-clad mountains to the sea, twenty miles away. Northwards and eastwards extended in glorious and shining array the magnificent chain of the Alps; glacier upon glacier, peak upon peak, range upon range of splendid mountains. Eastwards a fine rocky peak without a name and Mount Darwin, and looking south-westwards down the Tasman Glacier, from whence we had toiled our laborious way, the eye could follow the course of the great ice stream for twelve or thirteen miles, flanked by the grand mountains which sent down their tributary ice streams to join the mass in the valley below.

We gave three hearty cheers for her Majesty, and three for our proud little colony, and commenced the descent, going down backwards in the steps, and taking firm hold with our axes at every movement.

Time was precious, and on leaving the steps we ran down most of the less crevassed slopes, and soon found ourselves at the foot of the conquered mountain. Away we plodded down the glacier again—a hard, monotonous grind—till we arrived in the failing light at the system of crevasses on the outside of the turn of the glacier, close to our camp of the previous night.

This time we kept further out from the edge; but it was six of one and half a dozen of the other, for soon we were completely entrapped in a perfect maze of transverse and longitudinal crevasses, over which the only mode of progression was continued jumping.

This work in the dusk was anything but pleasant, yet had to be accomplished, and thanks to the aid of the rope, after leaping hundreds of them, we at length found our way off the side of the glacier to our tent.

How we watched the slowly warming 'billy' with eager eyes, and drank in fancy over and over again the pannikin of hot Liebig. How we shut the wind out and nursed the stinking kerosene stove! Alas for our hopes and our hungry stomachs, the lamp went wrong somehow, and the oil flowing over, the tent was on the verge of catching fire when Annan gave the whole concern a kick which sent flaming lamp, 'billy' and all outside. I hope the strong language and expressions of disgust have long since been forgiven us; but I really think they were justified.

Twelve hours' hard going did the Dome require. Von Lendenfeld took twenty-seven from the point of Malte Brun just opposite this camp.

Three hours' walking the next morning saw us back at our head-quarters, the Ball Glacier camp, where we found Dixon in active preparation for an assault on Aorangi, though not so strong as we could have wished.

Now a great council of war was held, the main point of discussion being as to whether we should attempt our long deferred ascent of Aorangi, which was, as usual, the chief object of our visit to the glaciers.

Here we were, with provisions for four or five days longer, the mountain apparently in good order, the weather perfection, and we were not pushed for time. The mountain had been inspected by various members of the party from different coigns of vantage. We had seen from a distance the *névé* fields leading on to the Linda Glacier.

Against this we had first to consider the state of Dixon's health. He was quite prepared, and anxious to try the ascent. We thought that it would be too much for him. Then there was the accident to the lamp, which was now useless, there was no firewood at the bivouac, 7,400 feet up, and no sure means of procuring water. Annan, too, had to leave to attend to his work down country, and I think, if the truth were told, that Johnson and I felt as if we had had enough of mountaineering for a time.

Yet we were very loth to turn our faces away again from the grim giant who had defied us so long, and it was only with much reluctance that we decided to abandon the project. So for the third time I retired from the ramparts of Aorangi unsuccessful, on this occasion without even so much as an attempt.

We came down to the Hermitage once more, and after a day or two's quiet rest yoked Dixon's celebrated tandem up, crossed the Tasman River, thus cutting off thirty miles of our homeward journey, and reached Fairlie Creek in two days. Here I took the train, whilst Dixon and Johnson drove home. The drive down and back—500 miles—was accomplished in twelve days' travelling with the same team of horses.

CHAPTER VII

FOURTH ATTEMPT TO CLIMB AORANGI

We reach the Great Plateau at last—Defeat again—The Crossing of the Ball Pass

'Perge et perage.'

ONCE again, on January 4, 1890, in company with Mr. Arthur Harper, a gentleman who had then done two seasons' climbing in Switzerland, I left Christchurch to try conclusions afresh with the monarch of the Southern Alps.

On this occasion we reached the Hermitage in two days from Christchurch, riding from Fairlie Creek, and crossing the Tasman River opposite Burnett's Mount Cook sheep station. Here we were joined by Annan, who had already conveyed the bulk of our impedimenta to the Ball Glacier camp.

On arrival at this point I at once remarked that the ice of the Ball Glacier had risen above its customary level, and seemed to be encroaching in a lateral direction—a circumstance which undoubtedly points to a cycle of advance in the great body of the ice, to be registered at the terminal face in years to come.

It will not be out of place here to give a description of our usual Alpine outfit, which may enable others to glean some idea of what is requisite and convenient for Alpine work in New Zealand.

The most necessary gear for ice and rock work is suitable boots, broad-soled and flat-heeled, shod well but not too thickly with heavy hobs, wrought nails being preferable to cast. An ice-axe for each man—not the light tourist's axe, but a guide's axe. Alpine rope is quite indispensable, and Buckingham's is the favourite make; we usually take two or three 50-feet lengths. Two tents, 6 feet by 8 feet and 6 feet by 7 feet, the former for use at the head camp, the latter a tent built after the 'Whymper' pattern with the floor sewn in, but capable of being pitched on inverted ice-axes lengthened by two 18-inch supplementary poles (an ingenious contrivance of Dixon's). Sleeping bags, 7 feet by 3 feet, made of blanketing, and covered with an outside bag of oiled calico, impervious to water. Aneroid, thermometer, prismatic compass, pocket compass. Goggles (neutral tint) are invaluable, and save the eyes from the awful glare which is always experienced on new snow and from the blinding sleet which drives in a storm. Folding lanterns (Austrian pattern) often enable one to find the way to camp when benighted or to make very early starts. A sheath-knife comes in very handy in camp, and a supply of fresh nails for our boots is never omitted, whilst a small 'Aurora' lamp stove is invaluable above the line of vegetation, and a shangai, or common schoolboy's window-breaker, is often useful in procuring birds for the *cuisine*.

For clothing, woollen shirts and knickerbockers of warm tweed material are the best, and great comfort is

Hochstetter Dome   Mount Darwin   Malte Brun Range

THE TASMAN GLACIER FROM NEAR THE DE LA BÉCHE CAMP

[From a Photograph by A. P. Harper

to be found in a loose-fitting boating 'sweater' worn over the waistcoat.

For provisions we generally rely on fresh mutton, to be fried in the pan or boiled in the 'billy,' bread, biscuits, rice, oatmeal, Liebig's Extract, chocolate, tea, and so on. A pound or two of fresh butter is always a boon, and a few tins of marmalade, whilst to some men onions supply the oft-felt want of a vegetable diet.

There is another indispensable, which here, as in the Caucasus, is very necessary. I refer to the late Mr. Donkin's naïve requisite at the end of his Caucasus list—'infinite patience'; and to this may be added fixedness of purpose, determination, and perseverance.

Mount Cook, or Aorangi, from a climber's point of view, is a very difficult peak to climb, even to a height of 9,000 feet, which our party attained on this occasion, chiefly on account of the length and tiresomeness of its approach. It is simply part of a great ridge which branches off in a southerly direction from the main divide of the Southern Alps. From its three peaks, all situated on this ridge, diverge four main spurs (or *arêtes*, as Alpine men call them). From the lowest and southernmost peak (11,787 feet) descends to the Ball Pass the southern *arête*; from the middle peak (12,173 feet) the eastern *arête*, descending on to an enormous buttress which separates the Ball and Hochstetter Glaciers; from the northernmost and highest peak (12,349 feet) two *arêtes* diverge, the north-eastern, separating the Hochstetter and Linda Glaciers and terminating in the ice of the Great

Plateau; and, lastly, the northern ridge, connecting with the main divide between Mount Tasman and St. David's Dome. A comparatively low rock saddle in this ridge occurs between the highest peak of Aorangi and the junction with the main divide, leading on one hand into the Linda Glacier, and on the other to the head of the Hooker Glacier. Aorangi is thus quite cut off from the west coast, and has, in fact, no 'western flanks,' as is generally supposed.

It was an intensely hot day, and scarcely a breath stirred as Harper, Annan, and I struck out on the now well-known route across the Ball and Hochstetter Glaciers for the Haast Ridge, but the clear mountain air seemed to rush into our lungs, putting health and strength into every fibre.

The mountains were glorious in the noonday glare, and the foliage on their lower slopes was in its gayest height of blossom. Now and then an avalanche would thunder down in the ice-fall or from the higher slopes above, or the whistle of a kaka down the valley could be detected. These and the merry tinkling of the surface streams were the only sounds to break the spell of silence and benignant peace which seemed to reign over all. These are the scenes which go straight to the heart of the true nature-loving mountaineer.

To reach the foot of the *couloir* by which three years previously Dixon, Inglis, and I had descended involved the usual amount of hot scrambling up *tali* or fans of detritus from the rocks above. Once in the *couloir* (which was snow-filled in places) we were not long in reaching our old bivouac, where we deposited our first batch of provisions, &c., our plan being to descend again that day and bring up more supplies on the morrow.

Coming down, Harper had an almost miraculous escape from swift and certain destruction. We were glissading on a snow slope when a mass of rocks broke suddenly away from above and whizzed down the slope at a terrific rate, passing within a few inches of Harper, who did not observe them coming, though both Annan and myself had seen the rocks start a hundred feet or so above him, and had shouted to warn him of their descent.

This was a warning to us to be careful how we trusted snow *couloirs* during the afternoon, after the sun's rays had done their daily work on the crust of the snow. It is by such lessons that we in New Zealand have learnt without the aid of Swiss guides to understand, to appreciate, and circumvent those dangers to which the Alpine climber is always more or less exposed.

Another fine morning saw us off again with sleeping-bags, tent, &c., and by noon we were up at the bivouac with three days' supplies. Only resting for an hour or two we pushed on upwards, intending to cross the Great Plateau—that ice-field of which we knew, but which we seemed fated never to reach—and find some sheltering rocks under Aorangi's uppermost slopes where we might spend the night.

In a few minutes we reached Mr. Green's sleeping-place, across which now lay a rock weighing some tons (another warning), illustrating forcibly the rotten state of the rocks.

We now roped and took to the snow, which led

first on to a small dip or saddle in the ridge (sloping off on the right to the Freshfield Glacier and on the left to the Hochstetter ice-fall), and then on to steep snow slopes leading up to the crest of the ridge overlooking the plateau, now about 1,000 feet above us.

We proceeded cautiously over many half-covered crevasses, and crossing the small dip or saddle took to the slopes beyond, now and then taking to the rocks on our left. The climbing was somewhat dangerous, mainly owing to the bad state of the snow, which would start away in avalanches, or give way on the edge of a crevasse just at the moment one put one's weight on to spring.

At length we gained the highest rocks, which proved very bad going and seemed likely to bring us to a stand; but the leading man going up the last fifty feet alone, sent down a spare rope, making one end fast above, by whose assistance the second man followed in safety, the last man making the swags fast to the rope below to be hauled up. In the act of hoisting them, however, one broke away, and commenced a furious flight down the slopes up which we had so laboriously toiled. To the swag was attached a pannikin and the tin cistern of our lamp stove, and at every bound we could hear the rattling of the tin as we watched the truant bundle leaping down, and we thought of what might be our fate, were it not for our trusty rope and axes, should we start unexpectedly down the steep slopes.

Still down went the swag, turning over on its ends and bounding over crevasses in a manner which made us quite envious. At last it hovered on a saddle. In breathless anxiety we wondered if it would stop, or whether it would take the slope to the Hochstetter ice-fall on the one hand, or the Freshfield on the other. One little effort more it appeared to make, and then away it went, careering down again towards the Freshfield ice-fall below.

Our hopes were shattered, and we were fast giving vent to expressions of despair when the career of the swag was suddenly cut short in a partially filled *bergschrund*, where it was brought up in some soft snow.

We dared not risk staying out for the night where we were without the lost swag, for no rocks affording any shelter were available, so determined, after making a little further progress to get a view of the plateau, to return to our bivouac at 7,400 feet—about 1,200 or 1,400 feet below our present altitude—and make a fresh attempt on the next day, weather permitting. The last man came up the rope, and we made our way up the final slopes of snow on to that great dome of glacier which we had so often gazed on from below.

Ah, what a sight burst upon our astonished eyes as we gained its summit!

It seemed the very acme of mountain glory in all the glories around us. A few hundred feet below lay that *terra incognita*, the Great Plateau, rounding off southwards to the Hochstetter ice-fall, bounded on the west by the giant form of Aorangi, on the north by Mount Tasman, and on the east by Mount Haast and the ridge of that mountain on which we now stood. The Linda Glacier could just be observed coming round the north-eastern *arête* of Aorangi, and on either side of it towered up to the heavens the two grandest mountains in New Zealand—Aorangi and Mount Tasman ; the former a lowering fortress of black rock and hanging glaciers, avalanche-streaked throughout, the latter an ice-clad mass with three summits, covered thickly with hanging glaciers overlapping one another as do shingles on a housetop, looking utterly unclimbable. Only two masses of rock are visible, over which avalanches constantly swept.

The sight is certainly the grandest of its kind I have seen in the Southern Alps, and Harper tried in vain to recall its equal in Switzerland.

After working our way upwards along the ridge to the nearest rocks we deposited a note of our visit in a pannikin, and building a small cairn over it, beat a retreat.

We experienced some difficulty in getting down the top rocks, but eventually gained our footsteps in the snow, and following down the route of the truant swag, recovered it from its snowy bed some 600 feet below the point where it commenced its downward journey.

We arrived at the bivouac just before dark, and had scarcely finished brewing a warm drink when down came a nor'-wester upon us.

Pitching the tent was out of the question, so piling stones upon it we spent a miserably cold night, and by the time morning came all thoughts of tackling Aorangi had flown, and soon we were speeding down to our refuge at the Ball Glacier camp again.

Thus ignominiously ended my fourth attempt to climb Mount Cook.

In the afternoon Annan went down the valley with directions to join us two days afterwards at the Hermitage, Harper and myself being determined to cross the southern spur of Aorangi at the head of the Ball Glacier, and work our way down the Hooker Glacier to the Hermitage.

THE FIRST CROSSING OF THE BALL PASS

Starting on a misty morning, we climbed what we call the Ball Glacier spur—a ridge which diverges from the main ridge of the Mount Cook Range at a point immediately south of the Ball Pass. It was by this ridge that Mr. Green's first and unsuccessful attempt was made, and up this same route I had climbed the previous season with the photographer.

The major part of the climb is easy, good foothold being obtained on the red sandstone rocks. In the upper part snow-fields alternate with the rocks. The Ball Glacier lies couched in the valley on the right, vast precipices going sheer down to it from the crest of the ridge, whilst the slopes on the left descend to the Tasman Valley.

After four hours of climbing we reached the main southern *arête*, and paused on the snow saddle for lunch and rest, and to admire the splendid prospect of the eastern faces of the mountain, and the ever-fresh, marvellous panorama of the Tasman Glacier.

Erecting a cairn on the rocks close by, and christening the saddle after that father of mountaineering—John Ball—we commenced the descent on a good snow slope towards the Hooker Glacier. All the mountains on the western side were enveloped in mist, which, however, fortunately hung high enough to enable us to discern the whole extent of the Mueller Glacier and most of that of the Hooker.

Bearing away southwards to avoid the crevassed parts of the slope below, we were soon enjoying a merry glissade—sometimes sitting, sometimes standing, whizzing down in a cloud of snow which curled up from our feet and showered down upon us.

Ah, the exhilaration of a good glissade! How you seem to fly through the air and cleave the fast-speeding surface! How the snow hisses and the axe grinds! How the excitement thrills you as you look out for danger ahead, or rushing avalanches behind! There is nothing to touch it—switchback railway, going downhill on a bicycle, skating—all are far behind.

In a quarter of an hour we entered a rocky gorge, and still down we sped on the snow, winding about in and out between magnificent rock precipices, until before another fifteen minutes had elapsed we emerged into the Hooker Valley, having come down 4,000 feet under half an hour.

Turning down the valley we kept to the old lateral moraine of the Hooker Glacier (which stands 235 feet above the present level of the glacier), and found it good walking

Once more, however, fortune forsook us, and an enemy in the shape of a south-west gale, accompanied with heavy rain, met us, against which at times we could scarcely make any headway. But struggling on we crossed the Hooker River on the ice of the Mueller Glacier, which at that time spanned it, and reached the Hermitage drenched to the skin at 4.30—eight hours from the Ball Glacier.

This was the first, and up to the time of writing is the only crossing of the Ball Pass, an excursion which ere long must become a favourite one, for a track is just completed to the Ball Glacier, where a two-roomed hut has been erected by the Government for the use of tourists and mountaineers.

A finer point of observation than the Ball Pass would be hard to find, as it commands the most comprehensive views of the Tasman, Hooker, and Mueller Glacier systems.

CHAPTER VIII

THE FIRST EXPLORATION OF THE MURCHISON GLACIER

Hard Swagging—Erroneous Maps—The Struggle for Starvation Saddle—Exhaustion and Hunger—Return

'Fresh fields and pastures new.'

I HAD often cast a longing eye in the direction of the Murchison Valley, and desired to explore those unvisited scenes which were as yet unknown and unseen by man. We had frequently during this visit to the district spoken of making an excursion in that direction should Mount Cook prove too heavy metal for us. Now was our chance, and we determined to take it.

Leaving the Hermitage with an addition to our party in the shape of Messrs. Wells, Timson, and Hamilton—the former two only intending to visit the Tasman Glacier, and the latter anxious to accompany us on the Murchison trip—we made the Ball Glacier camp, after the usual hard, hot grind over the moraine, by evening.

The next morning breaking fine, Wells and Timson went for an hour's excursion on to the glacier opposite, returning enchanted with the grand view of the Hochstetter ice-fall and the surrounding peaks, whilst the rest of us—viz. Harper, Hamilton, Annan, and myself—prepared swags for a two days' excursion up the

Murchison Valley, whose mouth could be discerned some two miles distant across the Tasman Glacier.

It is a fact worthy of notice that with the exception of mountaineering parties equipped for climbing—and the numbers of these could be counted on the fingers of both hands—Messrs. Wells and Timson were the first two tourists to venture so far up the Tasman as our camp, and since that time only one other has succeeded in reaching the same point, that gentleman being his Excellency the Earl of Onslow, Governor of the colony, whose practical penetration regarding all matters connected with New Zealand entitles him to the respect and gratitude of those subjects over whom he exercises vice-regal control.

Since the visit of Lord Onslow a track which had then been formed within two or three miles of the Ball Glacier has been completed, making the task of reaching the spot one of comparative ease and pleasure. Further conveniences for tourists and mountaineers in the shape of tracks and huts are now in course of construction by a far-seeing Government, who recognise the fact that New Zealand is fast becoming the playground of Australasia and the Switzerland of the South.

From careful inquiries made at the Survey Office, from Mr. Sealy—a gentleman whose early work of exploration amongst the New Zealand glaciers is too readily forgotten—and from the run-holders and station hands in the district, we had every reason to believe that the valley had only once been entered (by Mr. Burnett of Mount Cook sheep station), and that the face of the glacier had never been reached; only in one case could we hear of the clear ice having been seen—

viz., by a shepherd of Mr. Burnett's from a peak of the Liebig range.

There was therefore little or no doubt that we had a virgin field before us, and it was with feelings of intense eagerness that we pressed forward across the moraine-covered part of the Tasman Glacier, and up the shingle flats of the river-bed beyond, towards that massive, moraine-covered terminal face which fills the valley from side to side, five miles from the eastern lateral face of the Tasman Glacier.

The valley appeared to be a little over one mile in width. On either hand rose up most beautifully grassed slopes thickly covered with every variety of sub-Alpine foliage decked in the gayest height of blossom.

What a place for an artist's holiday! Flowers innumerable dotted amongst the richest shades of green—lilies, celmisias in great variety, Spaniards of many kinds with their golden and spiky heads of various shapes and sizes, from the orange-coloured dwarf to the great blue Spaniard with stalks occasionally ten feet in height; snow-grass with its graceful seed-stalks gently waving in the morning zephyr, which seemed to fan all Nature into a soft and dreamy repose—such wealth of colour, such variety of form, such grandeur of outline in the looming peaks above!

Yes; here the artist might fairly lose himself in delight amongst the subjects for his brush whilst drinking in the pure sympathy with Nature which seems to float in the very air.

It is no dream, this lovely valley, though it seems as one. But its flowers go with the warm geniality of summer, and when the cold winter comes round it dons its white garment of snow, hiding its beauties until the hand of gentle spring once more wakens them to burst forth anew in all their resplendent glory.

Proceeding up the valley between these magnificent mountains we kept moving onward in a north-easterly direction under the flanks of the Malte Brun Range, on to whose slopes we were now and then forced by encroaching streams from the meandering river, and we arrived early in the afternoon at a large boulder-fan issuing from a rocky gorge above, whence a magnificent waterfall descended. Here we boiled the 'billy' and lunched, making an inspection of the scene, which is one of the grandest beauty.

Far up in the heavens stands out a noble peak of the Malte Brun Range, rising out of a glacier which nestles in a basin of rock and bristles with *séracs* and pinnacles of blue ice pouring into the gorge below, from whence issues an imposing waterfall of seventy or eighty feet, sending up clouds of spray and drenching all within its immediate vicinity. From long action of the water an almost semicircular cylinder about ten feet in circumference has been worn into the solid rock, and the force of water descending this strange funnel seems to drive out in one direction a current of air which carries the spray with it.

All around this fall the vegetation is most luxuriant, and the rocks are covered with flowering plants in great profusion, and, in parts where the spray falls, plants, rare elsewhere, notably the myosotis, flourish in the abundant moisture.

Taking a more northerly direction we came to the terminal face of the glacier, which by aneroid measurement we made 3,640 feet—much the same altitude as our Ball Glacier camp The survey of the glacier has, however, since been effected, and this point determined as 3,305 feet.

The moraine is composed of unusually large polyhedral masses of rock, and is 200 feet in height at the main exit of the river, which is situated about the middle of the terminal face.

A backward view down the valley revealed but one distant peak—Mount Sealy—the northernmost of the Ben Ohau Range. This peak was evidently the only one from which the clear ice of the Murchison can be seen, if we except those of the Liebig and Malte Brun Ranges, and as none of these peaks have been ascended, this fact probably accounts for the Murchison Glacier, which is the second largest in New Zealand, having lain so long unexplored.

Proceeding up on the western side of the moraine, a new branch glacier descending from the Malte Brun Range opened out on our left, its lower ice forming a fine frozen cascade, whilst a waterfall of some 200 feet descends over a rocky face from its southern and hanging portion. To this glacier and fall we have given the name of 'Onslow,' in honour of his Excellency the Earl of Onslow.

As it was now getting dark we decided to bivouac for the night, and selecting a bed of small gravel amongst the larger stones of the moraine, we dined scantily on cold mutton and tea, and wriggling into our waterproof blanket-bags were soon ready for sleep. At first all our attempts at slumber were rendered futile by a congregation of keas, who hopped around within a few feet of us, jabbering and swearing in their own peculiar language at such a party of intruders on their domain.

The night was spent in comparative comfort, for we were beginning to feel the effects of our desperate swagging, and could go to sleep almost anywhere. It is simply astonishing what a man can put up with, when he has to; I have slept soundly in all sorts of queer positions, even upon a mixture of ice and sharp stones, without a tent and with only one thickness of blanket, when the thermometer has been several degrees below freezing point.

We were early aroused in the morning by the persistent attentions of the keas; they even went the length of pecking at our sleeping-bags, so tame and unaccustomed to man are they in these parts. We all wanted more rest, but it was not to be thought of if we adhered to our original plan of crossing a supposed saddle at the head of the Murchison to the Tasman Glacier by Mount Darwin, and returning to our head-quarters after accomplishing the circuit of the Malte Brun Range.

We were soon off, and toiled up the small valley formed by the lateral moraine of the glacier and the slopes of the Malte Brun Range. About a mile or so up we observed another glacier lying in a comparatively low saddle above us on our left, beyond this a rocky spur, and then another and larger branch glacier which for a time we took to be the main body of the Murchison, as indicated by the maps. We made for it and climbed its enormous face of ice, and then we discovered our error, for there, a mile away across the moraine, lay the clear ice of the Murchison, and far, far away northwards, the valley extended completely filled with a magnificent *mer de glace* of pure white ice. We stood transfixed, for none of us had imagined that such a grand glacier lay beyond.

Now we saw what was before us, and for a long time debated as to our ability to face the work ahead.

Hamilton was shockingly out of condition, and a sinew in my leg was becoming painful, Nature at last rebelling against the strain to which she was being subjected. We had a very scanty supply of provisions, and evidently it meant spending another night out if we proceeded.

The temptation was too much for us. We could not let this prize slip through our fingers, so we decided to go on and put ourselves on starvation rations rather than turn. Away we struck over the moraine, and in an hour's time reached the clear ice, here much crevassed. Crossing with some difficulty we lunched on the eastern side. Casting our eyes backward we could see splendidly all the fine peaks we had been passing under, and could observe the continuation of the range north-eastwards with five or six more branch glaciers, the final one northward leading to a snowfield with a saddle at its head. This, then, must be our saddle, we thought. But it seemed hopeless to cross it in our tired condition and with our heavy swags.

We set our teeth, however, and went doggedly forward, striking out on to the clear ice again and making a north-easterly course, at each step realising more and more the grandeur of the immense ice-field now gradually opening out and unfolding the wealth of mountain glory which encloses it.

We tried in vain to identify Mount Darwin or the most northerly peaks of the Malte Brun Range, which we knew were amongst those on our left, and, according to our reckoning by the maps—framed from Von Haast's—which seems to have been compiled from guesswork as far as this locality is concerned—we should at this time have been on the Classen Glacier, which lies at the southern head of the Godley River, and, in reality, was some miles north over the Liebig Range.

Passing several branch glaciers on our left, and observing that those on our right were assuming larger proportions, we sidled obliquely across and made for the snow-field leading to the saddle which we had every reason to believe led into the Tasman. Altering our course to due north, and crossing the lower and sloppy part of the snow-field, which was flat and quite undrained by crevasses, we were soon on snow in miserable order, and putting on the rope we wound our way gently upwards amongst the crevasses now beginning to appear.

We had just six hours of daylight, and considered we could reach the saddle in four if all went well, which would leave us two hours to find a bivouac on the other side, provided the descent were feasible.

We found it necessary to change leaders again and again to distribute the arduous task of breaking steps in treacherous snow, just in the condition to let us through knee-deep as we put our weight on it, and we had to observe the greatest caution in crossing the crevasses, which were very deep and almost invariably half covered, or had edges fringed with cornices of soft snow, which at times had to be removed or trodden down to enable us to obtain a sound footing on the hard edges concealed beneath it.

The grade steepened, and we all felt the hard work, more especially Hamilton, who was sadly out of form, but stuck to his work like a Trojan, despite the cruel punishing his swags were giving him.

Now we had to make our way across a slope where an avalanche had recently come, and, worse than all, a thick mist accompanied by a keen wind began to come over our saddle.

Still we pushed slowly upwards, resting every few minutes. Thoughts of turning began to arise in our doubting minds. But this would not do with the *col* so nearly within our grasp, and the cry was almost one of 'Death or victory!' as we plodded laboriously upwards. Sometimes we could not see fifty feet ahead, and were compelled to steer by the compass, taking bearings of crevasses and ice-blocks as we proceeded. Now and then the mist would lift for a moment and we could catch a glimpse of the longed-for saddle, and at last, when within a couple of hundred feet, Annan and I cast off on a separate rope, made a rush—as much of a rush as we could muster up—for the goal, hoping at least to get a glimpse of the other side ere the mist became too dense.

Hurrah! the saddle was conquered! But what lay beneath? Mist! Mist! Nothing but a thick impenetrable mist.

The other men arrived, and simultaneously, as if by some providential magic, the fog began to dissipate.

As it cleared we looked in vain for the familiar points at the head of the Tasman, which Annan and I knew full well. 'Where's Darwin? Where's Elie de Beaumont? Where's the Dome?' No point in sight could be associated with the prominent features of the Tasman. As the low-lying portions of the mist disappeared, we observed that the glacier below flowed to the right! The Tasman should have flowed in the opposite direction

The truth flashed upon us, and a great cry of surprise went up, 'The Murchison! The Murchison!' The very glacier whose middle parts we had left three hours previously.

Then, leaving Hamilton exhausted on the saddle, the rest of us struck up to some rocks 300 feet higher on the right, and once more a great shout arose as Annan and I saw coming into view the unmistakable double top of the great Hochstetter Dome, whose proud summit we had trodden the previous season.

From these rocks we observed that the course of the glacier commenced under a peak on our left (which must be Mount Darwin itself), and running in a northerly direction for some four or five miles, turning round the end of the spur upon which our saddle was situated, assumed a south-westerly course.

The true saddle between the Murchison and Tasman lay across the glacier below, north-west. Straight ahead of us, north by west, visible over a rocky and unnamed peak on the opposite side of the valley, lay the Dome, then to the north another snow saddle, evidently leading into the Whymper Glacier, and so on to the Wataroa River of the west coast. Following round the range to the right a very fine mountain stands boldly up; to the right of this, again, is situated yet another snow saddle, which we concluded must lead into the Classen Glacier.

The result of the Government survey of the Murchison Glacier, just completed (1891), confirms our surmises regarding the topography of this interesting district.

We were astonished at the great length which the Liebig Range assumes, for it bounds the glacier throughout the whole of its eastern side, diverging from the main chain of the Southern Alps some distance north of the Hochstetter Dome.

Any attempt at a description of the panorama from our saddle would be useless to convey an adequate idea of the view. Harper classed it as similar in character to the views obtained at high altitudes in the Bernese Oberland. An aneroid reading gave our height as about 7,900 feet, but this was much out, as by the recent survey the height of the saddle has been trigonometrically determined as 7,194 feet. Our estimate of the length of the glacier at the time was twelve miles, and the survey has now fixed it at eleven and a half, whilst the average width is as nearly as possible one mile.

Reaching the saddle into the Tasman was now quite out of the question, for it would involve a descent to the valley below, the crossing of the upper parts of the glacier, and the scaling of more snow slopes, which appeared to us to be impracticable owing to the

MOUNTAIN LILIES (*Ranunculus Lyalii*) [*Wheeler & Son, Photo.*

numerous crevasses. In addition to this, one man was lying *hors de combat* on the snow suffering from exhaustion and vomiting. Evidently the only course we could pursue was to retrace our upward route, and that as quickly as possible, for there were but three hours of daylight left to reach a bivouac in the rocks lower down.

After erecting a small cairn, depositing a record of our ascent, and giving three cheers for nobody quite knew what, we roped up and began the descent.

It is astonishing how one's spirits revive when a fresh set of muscles is brought into action, aided by the force of gravitation, and though we had been defeated in our attempt to reach the Tasman, what did that matter? Though we were half-dead with starvation—'Starvation Saddle' is now the name of our *col*—and though a real weariness of the flesh had taken hold of us, what matter? We had explored (I might almost say discovered) the great glacier we had come out to see, and would be able to settle all sorts of topographical errors in the maps, and could speak with authority about many square miles of Alpine country hitherto entirely unknown.

Our spirits rose as we descended, despite our hungry and tired state, and we once more wound our way down among the crevasses, and reaching the glacier again made for the lowest point we could before night closed in. But we had an hour's cruel moraine work in the dark ere we found a sleeping-place on a bed of lilies, where we boiled our last drop of Liebig and divided our remaining crust of bread.

It rained a little during the night, but we did not

care for that with our oilskin bags, and sleep visited our weary eyelids as it had never done before.

Hamilton's condition had improved, but his feet were sore and he was very weak when at 4.30 A.M. we once more set off for our home on the glaciers—the Ball Glacier camp. The prospect of boiled rice and fresh chops lured us on as we made our way down the valley, and putting forth our last remaining energy we made the ever-welcome refuge in eight hours, Harper, who had most left in him, going on ahead and preparing a substantial feed for the stragglers behind.

Oh, that tin plate of rice, *and* those chops, *and* that tea!

Now came an exhibition of pluck rarely seen. After two hours' rest Hamilton said he must reach the Hermitage that night; despite our dissuasions he determined to go on, and Annan generously volunteered to accompany him. These two men actually reached the Hermitage that evening at 8.30. It was the pluckiest day's work I have ever seen done in the mountains.

Harper and I came down next day in a snow storm, with fifty-pound swags.

Many people seem to think that a visit to the Alpine regions necessarily entails contact with very cold weather, even in the summer time. This is quite an erroneous idea, for on this occasion the thermometer readings at the lower camp varied from 42° Fahr. in the morning to 72° in the evening, and I should think that even during the coldest night the instrument did not register much lower than the first-named figure. We frequently went about in shirt and knickers only, and the usual complaint is of the heat, not of the cold. Some men suffer a good deal of discomfort from sunburn. I myself am a victim in this respect. It is the upper and freshly fallen snow which is so ruinous to the epidermis, the reflection from the new and unmelted crystals being so great as to cause the skin to assume a dark chocolate colour even during one day's work amongst it.

Sometimes blisters form, after which the skin puckers up and eventually peels off in patches. The noses of persons possessed of aquiline features are usually a study in themselves after a day or two's exposure on new snow.

CHAPTER IX

FIFTH ATTEMPT TO CLIMB AORANGI

Avalanches—The bivouac again—First attempt repulsed—Second attempt—The Great Plateau—The Linda Glacier—Hard work step-cutting—The terrible couloirs—Victory at last—Descent by lantern-light—Back to civilisation

> Are not the mountains, waves, and skies a part
> Of me and of my soul, as I of them?
> Is not the love of these deep in my heart
> With a pure passion?

WHYMPER was eight seasons climbing the Matterhorn. Dent made innumerable attempts ere he conquered the Aiguille du Dru—why should we despair about Aorangi?

We certainly were at a great disadvantage as compared with Swiss mountaineers; we had to begin at the very bottom rung of the ladder, having no trained guides. But I am confident that if we had been as many years climbing with guides as we have been without them we should be far less proficient mountaineers.

Probably our case is a unique one, and I doubt if there exists another instance where two or three novices —at any rate at ice work—have banded themselves together and gone systematically into heavy Alpine work 'right away' (as the Americans say), doing all

Starvation Saddle

[Wheeler & Son, Photo.

LOOKING ACROSS THE MURCHISON GLACIER

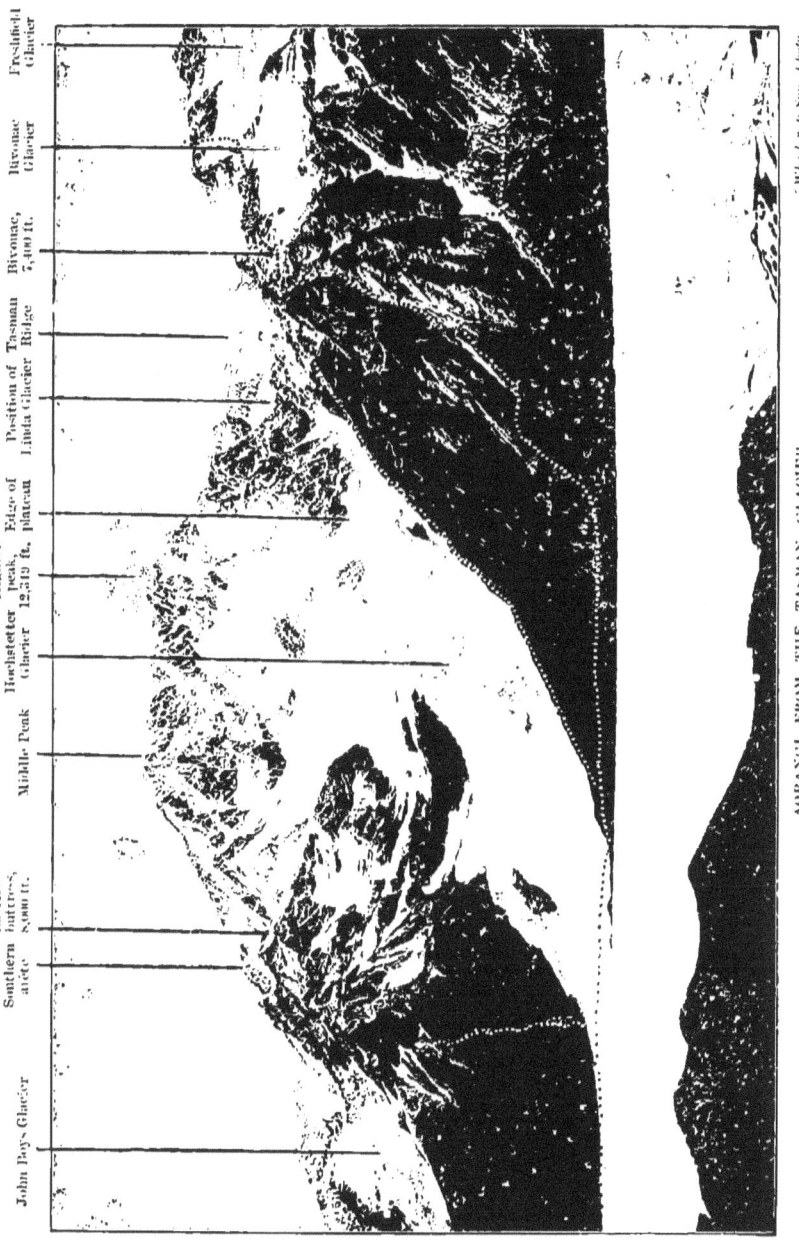

AORANGI FROM THE TASMAN GLACIER

their own porterage and guide work from the start. We learned fast from that best of masters—'hard experience.'

Had we been consistently following in the footsteps of trained guides we should not have concerned ourselves about this, that, and the other, but would have left everything to the men of experience, simply being towed about in their wake; whereas we have been obliged to train and exercise all those qualities which a guide possesses, perforce.

Naturally, too, a colonial life is more calculated to teach self-reliance and independence, and from our earliest schoolboy days we have been accustomed to rough work on the hills, pig-hunting, &c., and in camp life on all sorts of hare-brained expeditions. I have, indeed, been in many an awkward place amongst rocks when out on the foot-hills, and must have—perhaps unconsciously—acquired many of those qualities which denote the cragsman.

Want of fixedness of purpose had often lured us away from the peak, and temptations in glacier expeditions had thwarted our determination. I felt confident, however, if Dixon and I got together again we should make a good fight of it with the mountain, for we had learnt to place confidence in each other in many rough trips, and Dixon was a man after my own heart for determination.

On December 1, 1890, then, for the last time Dixon and I found ourselves on the way to the Mount Cook district; we reached Burke's Pass that evening in an express waggon which contained besides ourselves two small Rob Roy canoes, it being our intention to navigate the Waitaki River from Aorangi to the sea—should we not previously leave our lifeless bodies at the foot of some precipice or frozen in a crevasse, as many of our friends prophesied.

December 2 saw us crossing the Tasman River in our canoes ten miles below its exit from the glacier, and as it was in flood and running full ten knots, with waves four or five feet high in the rapids, we had an exciting time of it, yet managed to reach the Hermitage side in safety, but not without shipping a good deal of water. This was the first case of a boat of any kind being on these rushing waters, and our good friends in all directions prophesied dire disaster to what they were pleased to term our 'rash venture.' We are getting quite used to these consolations of our friends, who seem quite disappointed that we do not afford them some sensational obituary matter in the daily papers.

Again the faithful Annan was at hand, and greeted us at the Hooker wire rope with the pleasing intelligence that our camp at the Ball Glacier was fixed and our swags conveyed there. The Government surveyor (Mr. Brodrick) and his party were at hand too, and working their way to the Murchison Glacier to make a survey in continuation of their work on the Tasman; we spent the following night in comfort at their lower camp, one mile above the terminal face of the Tasman Glacier, to which point a horse track had already been formed through the scrub.

Again we carried our swags up that cruel piece of walking to the Ball Glacier camp, stopping half-way for lunch at our customary resting-place—'The Cove' —a snug little nook in a rock-face where a rill from

Mount Cooper   Classen Saddle   Harper Glacier

THE MURCHISON GLACIER

[*Wheeler & Son, Photo.*

the mountain-side offers cool refreshment to the weary swagger.

Friday morning, December 5, found us early astir, and making up swags of blanket-bags, tent, tinned meats, biscuits, chocolate, raisins, prunes, rice, oatmeal, Liebig's Extract, and all such necessaries as might ensure sustenance and a certain degree of comfort at a high bivouac. Seeing that our boots were well nailed, our ice-axes and snow-goggles in good order, we struck out across the Ball and Hochstetter Glaciers and reached the foot of our climb—the southern termination of the ridge of Mount Haast. Here we deposited a small supply of provisions as a standby, in case we should be driven back by bad weather or by some unforeseen cause.

The day was very warm, and as we toiled slowly up under the weight of our heavy swags (we were carrying enough provisions to last us for some days) the perspiration streamed from every pore, and the sun's rays seemed to penetrate with singular fierceness.

Soon we came to the lower termination of the new and unmelted winter snow in the *couloirs* or ditches between the rock ridges, and as the day advanced the hissing avalanches came down these slopes with increasing frequency, and falling stones and rocks now and again passed close by us. The snow being in such a loose and slushy condition it was imperative that we should avoid it as much as possible, but climb as we would we could not help occasionally crossing a snow-filled *couloir*, and this had to be accomplished with much celerity and caution.

Annan was particularly anxious concerning the

'shocking state of repair' of these lower slopes, and seemed to lose his nerve entirely, though he is accustomed to work on the higher beats in mustering, &c., and he declared his intention of going no farther than the bivouac at 7,400 feet, which we reached in the afternoon. We at once saw that it would be useless and dangerous to persuade him to join us in the final assault, for if his self-reliance failed on these lower slopes, what would the state of his nerve be on the upper ice work where so much step-cutting would be necessary? Dixon and I knew that we ought not to try to ascend the peak alone, that such work as we—two guideless amateurs—were about to attempt, would not be looked upon with favour by such a body as the English Alpine Club; but we were so tired of knuckling under to Aorangi that we were becoming desperate, and we decided to try conclusions without a third man.

Two hours of excavation work removed two feet of snow and eighteen inches of ice from our bivouac, revealing the faithful 'Aurora' stove and sundry potted meats left twelve months before by Harper and myself, and soon we had the tent pitched and were snug for the night.

At three o'clock on the Saturday morning Dixon and I crawled out of our sleeping-bags, and by 4 A.M. we were on the snow slopes, determined to make a vigorous attack upon the peak which had so long defied us.

Two hours on fairly good snow slopes and a scramble over a nasty slab-like face of rock, and once again the plateau, and that glorious scene of Aorangi and Tasman, were before us.

But the wind had risen quickly and was blowing a gale from the south-west—the cold quarter. To face such a wind for any length of time, or to attempt to climb Aorangi against it, would be simple madness, so we turned and ignominiously fled to the refuge of our bivouac, 1,200 feet below, which we reached at seven o'clock, having been but three hours absent.

We then sent Annan down, as we were keeping him from his work in the lower country, telling him to leave word with the survey party that if we did not arrive back at the Ball Glacier by Monday night something would probably have gone amiss with us.

During the day the gale blew itself out, and next morning at 3.45 we were in our steps of the day before, reaching the plateau in an hour and a half. The morning sun lit up the peaks with a rosy glow, soon his piercing beams forced us to put on the goggles, while the crust of the snow began to soften under the great power of penetration which the rays possess in the rarefied air. This forced us to plod onward in slushy snow as we headed right for the Linda Glacier, which we could see rounding the point of the north-eastern *arête* of our mountain.

On our right rose Mount Tasman clothed in ice, from which during the night an immense avalanche had descended. We walked close to its furthest point of motion as it lay stretched out on the level snow-field like a gigantic breakwater, and found it to be 300 paces in width; Dixon estimated that it covered from forty to fifty acres.

We now put on the rope, as crevasses began to appear in the gently rising slopes to the Linda Glacier.

On our left we thought that the north-eastern ridge looked practicable, but deemed it better to rely on a route chosen by so able a mountaineer as Ulrich Kaufmann, and kept on our course for the Linda Glacier, taking ten-minute spells at leading and breaking steps in the soft and slushy snow, and winding our way amongst ever-increasing crevasses in search of snow bridges over which we would cautiously crawl.

Now we would have a stretch of gently rising snow, then a crevasse or perhaps a *bergschrund*, followed by a steep ascent for 100 or 200 feet, then a divergence to one side or the other to avoid a chaos of *séracs* or blocks of tumbled and broken ice; and so on, hour after hour. About noon we had gained a considerable elevation above the plateau and were well round the corner on the Linda Glacier. Into this elevated valley the sun poured down through a rarefied atmosphere on to slopes on either hand which reflected all the light and heat. The glare was something dreadful, and before midday our faces and hands had assumed the customary chocolate colour, and the skin was literally broiled off me; Dixon did not suffer to such an extent. The heat was most intense, though not of the enervating kind which one feels at lower altitudes.

Viewed from this quarter Aorangi presents a totally different form than from any other, and we began to be sanguine about accomplishing our task. I was in possession of notes and sketches of the route kindly sent me by Mr. Green, and these were of material assistance to us.

Before us lay the final peak with its capping of ice. From the summit, now in full view, descended in a north-westerly direction to the right a steep rocky *arête* connecting with the ridge leading on to Mount Tasman. From the lower parts of these rocks steep ice slopes streaked with marks from falling rocks descend to the upper portions of the Linda Glacier, bounded all along their lower termination by an immense *bergschrund* which severs them from immediate contact with the glacier itself.

On the left of the summit slopes the north-eastern *arête*, consisting of a ridge of alternate knife-edges of ice and *gensdarmes* or towers of rock. The northern side or face of this ridge descending to the Linda Glacier is composed of very steep slopes of ice set with three immense masses of red-sandstone rocks, with two ice-filled *couloirs* or ditches between them. Up these two *couloirs* lay our route. We thought, however, that by leaving the glacier and taking to the crest of the ridge we could improve on the route, but soon found that the change was a mistake, and so struck back on to our old course up the middle of the glacier, the final slopes of which were very steep and exposed to the chance of avalanches from either hand.

It seemed a hopeless task this plunging through soft snow hour after hour, and it was nearly one o'clock ere we gained the edge of the big *bergschrund* and with difficulty discovered a sound enough snow bridge. Shortly before this an incident occurred in crossing one of these snow bridges which brought forcibly before our minds the serious nature of the work in which we were engaged. I—the lighter man by two stone—had crawled over in safety, and planting myself well in the soft snow above, was taking in the slack of the rope as Dixon followed, when suddenly he went through up to his armpits and was dangling in space, held up by a thin crust of snow and by the rope from above. I pulled with the strength of despair, and Dixon struggled till he secured a hold somehow on the other lip of the crevasse and got out.

That sort of thing is all very well to look back upon and talk over afterwards, but I am not likely to forget for many a long day the sensation of holding up a thirteen-stone man under such circumstances, and I must say that I should have been much easier in my mind if we had had such a man as Emil Boss or Ulrich Kaufmann on one end of the rope.

Immediately after crossing the big *bergschrund* step-cutting commenced; and from this point upwards every step, other than those on rocks, had to be cut in hard ice.

It is no easy task after climbing steadily for nine hours in soft snow to set to work and cut steps, especially when one knows that a slip must on no account be made, for with two men only on the rope it would mean a sudden descent to the crevasses or precipices (as the case may be) below, and our certain destruction.

An hour's steady work and we gained the foot of the lowest rocks, which were found to be quite unscalable. We then sidled round the base of these rocks to the left and commenced cutting steps up the first *couloir*, keeping close into the rocks on our right, on which we could get an occasional hand-grip. Ice blocks were continually coming down from the broken masses overhanging the top of the *couloir*, but luckily none struck us. The descent of an ice block in such steep ice slopes is something to remember. First a rattle above, and then 'swish, swish' as the first leaps begin, followed by a 'whir-r-r-r' and a 'hum-m-m-m' as, like a flash of light, a spinning and ricochetting object goes by and is lost to sight over the brink of the precipice below, or perchance is detected spending its momentum on the soft snow slopes 1,000 feet down.

These falls of ice on the upper slopes are not like the hissing avalanches, which sometimes even *crawl* down the lower snow slopes, but come down with the speed of light, and are calculated to strike terror into the heart of the stoutest-nerved climber.

We crossed the *couloir* near its head, partly on ice and partly on rocks, amid the gravest peril from showers of ice, and took to the rocks on our left, which were both dangerous and difficult, mainly owing to their being here and there coated with ice. Soon they became quite inaccessible, and we were again forced towards our left on to the ice slopes in the second *couloir*, and here we found the ice even harder, and we could only make an impression on it with the spike end of our axes. To add to the difficulty, the angle of ascent became steeper, inclining in places to about 60° from the horizontal.

We negotiated this *couloir* in a similar manner to that below, but water trickling from the overhanging rocks formed awkward hummocks of ice on the slope close to the rocks, over which we thought it almost impossible to climb, and to go out into the middle of the *couloir* was impossible, owing to falling ice.

Time was quickly passing, and we had a terrible

fight to reach the head of the *couloir*. The rocks now shaded us from the sun's rays, and soon our hats, coats, and the rope were frozen as stiff as boards, while the cold was so intense as to cause the skin of our hands to adhere to the steel of the ice-axes.

It seemed now more than ever a hopeless task to reach the final ice-cap, which we knew could not be far above us; but we silently and doggedly cut away, and at length were rewarded by finding the rocks on our right practicable; taking to them we were soon on their crest, and the ice-cap of the mountain lay straight before us. An easy bit of rock-climbing led up to the slopes, which we found to be covered with a peculiar form of lumpy and frozen drifted snow. At the top of the rocks we looked around in vain for Mr. Green's cairn, with his handkerchief and Kaufmann's match-box, left on the occasion of their ascent in March 1882. Doubtless they have either been long since swept away by falling ice or were buried in the terminal of the ice slope, which in December would encroach farther down upon the rocks than in March.

Dixon now counselled a retreat, arguing that we had virtually overcome all the difficulties and had only the final and easy slope to cut up; but I persuaded him to stay a little longer and make a push for it, or at least as much of a push as we were capable of making.

It was half-past five. Four hours and a half we had been toiling from the head of the Linda Glacier, thirteen hours and a half from our bivouac, without any halt to speak of. A wind began to blow from the north-west, adding fresh cause for anxiety about the descent. One thing was certain—if we wanted to get

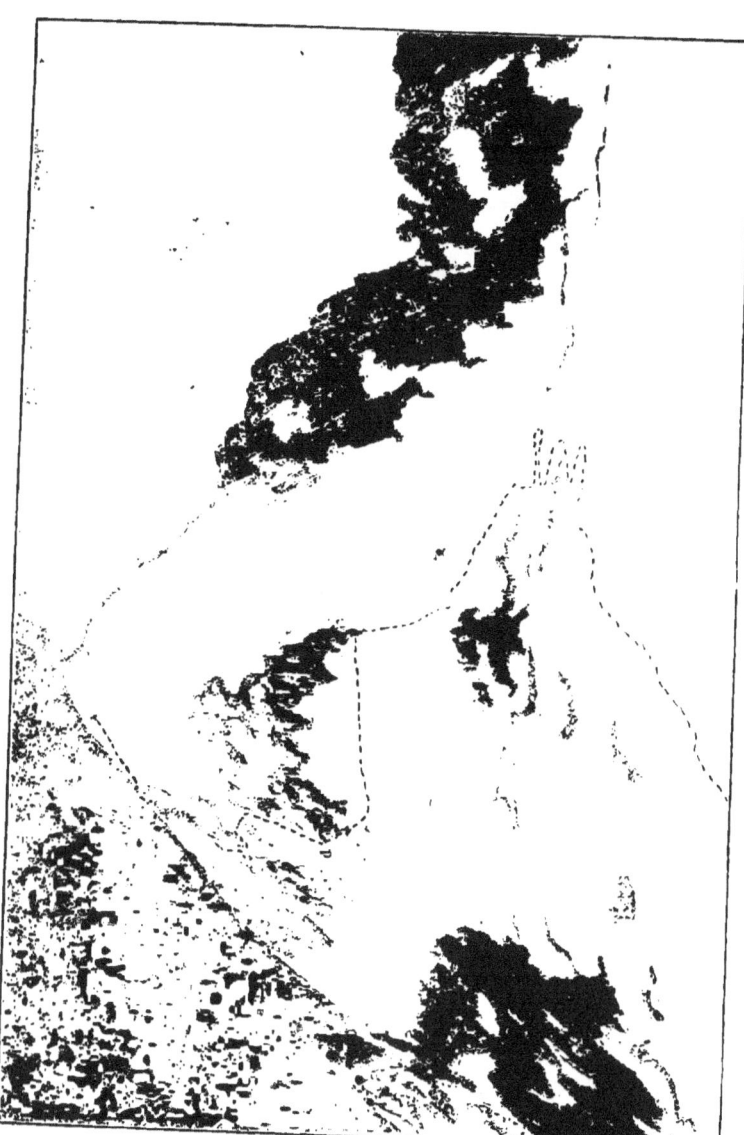

AORANGI: THE HIGHEST PEAK

[*From a Water-colour Sketch*

down alive we should have to reach the Linda Glacier again before dark.

We worked as hard as we were able at step cutting for another fifteen minutes, but only made slow progress; yet there was the cornice, just away to the right, the crest of the ridge to the left, and the top scarcely a stone's throw above, with no difficulty in the way. What would we not have given for another hour of daylight? How could we turn away when so near to a complete victory over our old foe?

Dixon again suggested turning, and I could not do otherwise than defer to his advice, for already we were caught in a trap, and should bad weather come upon us—and the wind and cold were fast increasing—before we reached the Linda Glacier again the probabilities were that we never should have returned from the giddy heights of the great Aorangi, the 'Sky-piercer.'

The height of the mountain is 12,349 feet; our aneroid read at our turning-point 12,300, and we reckoned the summit to be 140 feet above us. The slight error in the reading of the instrument would be accounted for by the impending change of weather.

The view is magnificently comprehensive. Looking northwards we could see clear over the top of our giant neighbour, Mount Tasman (11,475 feet). On the western side, the ocean, but twenty miles distant, was covered by a mantle of low-lying clouds creeping into the bays and inlets of the coast, studded here and there with islanded hill-tops, and stretching away to what seemed a limitless horizon on the west. A streak of blue ocean showed through the cloud mantle near Hokitika, seventy miles northwards.

North-eastwards the glorious array of the Southern Alps extended, presenting a panorama of such magnificence and comprehensiveness that it defies any attempt at description. It is one of those vast pictures which are indelibly impressed upon the memory—one of those overpowering examples of Nature's sublimity which seem to move a man's very soul and call him to a sense of his own littleness.

Close under us lay the scenes of all our joys and sorrows of the past five years: the Tasman Glacier, encircled by those splendid peaks and snow-fields whose forms we had learned to know and love so well; further afield lay the Liebig Range, and, showing over this, Mount Jukes and his attendant satellites of rocky peaks. Beyond this again, far, far away in the blue and indefinite east, we could distinguish the hills of Banks Peninsula, close to our homes near Christchurch, whilst we could imagine that the blue haze distinguishable there was indeed the eastern ocean, 120 miles distant.

It will, of course, be said that we did not make the complete ascent of the mountain. Be that so; neither does Mr. Green claim that honour, though for all practical purposes to be on the ice-cap of Aorangi means the same thing as being on the top. Mr. Green's highest point must, according to his sketches, have been as nearly as possible 100 feet above ours.

But we could not spare time to moralise and rest as we should like to have done, for it was imperative that the terrible ice slopes should be descended before the light failed, and at a few minutes to six we began to go down backwards in our steps, taking a firm hold with our axes at every step.

This going down is a fearful strain on the nerves, and requires the greatest steadiness and caution. In hurrying down the easy rocks we missed a mark on a snow patch which Dixon had made to denote the right route, and this mistake at the outset caused us nearly half an hour's delay before we found the right spot from which to leave the crest of the rocks. Dixon led down the rocks and I followed, every now and then taking a turn round any prominent projection with the rope and easing him down, whilst he in turn secured a good hold and took in the slack as I came down.

Bad as it had been coming up the top *couloir*, it was infinitely worse going down, for what was trickling water on the upward journey was now solid ice, and many of the steps were filled with re-frozen chips of ice from the steps we had cut above, and these had to be cleaned out before we could get a secure foothold.

Cutting steps *up* is one thing, and cutting them *down* another, for on a steep slope one cannot turn round face downwards to get at one's work, which in the case of going up-hill lies convenient to the hand.

How we did get down without the fatal slip which I was momentarily expecting would be made by one or the other of us I never could quite understand.

The rocks below the topmost *couloir* were negotiated and the lower *couloir* reached. This was not so difficult to descend, and the effect of the frost was such as to prevent such a continual shower of ice blocks from above, thus minimising one prominent danger.

The lower parts of the *couloir* were reached, and here are situated the rocks which form the ledge upon which with Boss and Kaufmann Mr. Green stood out for the night. There are several ledges accessible, but Mr. Green's party must have been upon one of the higher, for on some of the lower ledges there is room for a dozen men to stand or even lie down, though scarcely space enough for a circus or Wild West show, as Dixon humorously suggested. The light was now fast failing, and we strained every nerve to reach the big *bergschrund* below before darkness was upon us.

We were just in time and that was all, and the frail snow bridge was passed by our sliding over on our backs; I, the lighter man, led, and Dixon followed as steady as a rock—not a Mount Cook Rock, but the proverbial one.

We had now been seventeen hours with every nerve and muscle constantly in action, and yet, as the darkness set in and the awful glare of the sun had left us, we began to freshen up, and lighting one of our Austrian climbing-lanterns we retraced our footsteps of the morning, being most careful never to deviate from them. Soon it became very dark, for there was no moon, and we could but dimly distinguish the ghostly forms of the white-robed peaks which shut us in on all hands.

Hour after hour we plodded on. On one occasion we were brought up by the crevasse into which Dixon had nearly fallen in the morning; it had opened wider during the day, and only after walking along its line of fracture in both directions for half an hour did we discover a bridge which seemed sufficiently strong. We crossed in our usual way, sliding over at full length, and putting some weight on to our axe-handles laid lengthways on the snow to distribute the weight as much as possible.

As the night wore on, the crust of the snow became harder, and after passing through that most unpleasant crusted stage when it will bear until all the weight is put on one foot, became quite pleasant to walk upon, and over the lower part of the Linda Glacier and across the plateau we made a fair pace. As we reached the rise off the plateau on to the Haast Ridge the wind increased in violence, and we had great difficulty in keeping our lanterns (two of which we now kept going) alight.

The crest of the ridge was gained, and the descent of the dangerous snow slopes to the bivouac, 1,200 or 1,400 feet below, commenced. We were soon in trouble again amongst *bergschrunds* and crevasses, and on two occasions, in going down and feeling for the next step behind, I found on showing a light that my *hind* leg was dangling in a crevasse!

I must not weary you, dear reader, with further monotonous descriptions of crossing these deadly enemies of the mountaineer, suffice it to say that after an exasperating hunt on the steep slopes and in the dark for our bivouac—the candles being just finished— we finally discovered it at 2.45 A.M., an hour before daylight, having been twenty-three hours constantly hard at work without any halt worthy the name.

Sleeping soundly till 9 A.M. we made up our swags, and by 11 A.M. were on the downward route again for the Ball Glacier camp.

It was quite a wrench to leave our friendly rock, which had become a haven of rest and refuge to us on this upper beat. Five nights have I spent under its protection at different times, and as often have I arisen with the early morn to gaze upon those vast and sublime solitudes of Nature so grandly unfolded to view. From this little home—out of which if one stepped one had to be careful not to lose one's footing and make a rapid descent to the Hochstetter Glacier on one hand or to the Freshfield on the other—I have seen the rosy tints of the newly-born day creep downwards from the hoary snow-caps of the mountains, and when evening drew on have watched the afterglow envelop all in its warm embrace, even black rocks turning to a deep crimson which seemed to pervade the higher peaks ere the black and cold night once again grasps them in his icy hold.

Here had tired limbs been laid to rest whilst wearied minds dreamed dreams of success and hope, gaining renewed vigour with the morning light to go forth afresh into new struggles and enjoyments. Here in the heart of great Nature's solitudes the thoughts flew back to homes of comfort and of love. What wonder that we should have formed associations with such a spot?

The Ball Glacier camp was reached at 4.30 P.M., where we found Mr. Sladden of the Survey party anxiously awaiting our arrival, with that forethought which shows the kindly feeling and consideration for others that characterises men of worth in these outlandish parts.

That evening Dixon went across with Sladden to the Survey camp in the Murchison Valley, leaving me to wait for an expected friend from Christchurch.

Here I was quite alone amongst the mountains,

with plenty of time to muse over the events of the past few days and to let my wandering thoughts fly back even further, to the struggles of the past five years whilst attempting to conquer Aorangi.

What is the climber's reward for all his trouble? Why does he climb? Who can tell?

Is it renown he struggles for? No; I am convinced that is a very infinitesimal motive. For mercenary ends? No; there is no financial harvest to reap.

I have often tried to think why men undergo such labour and hardship, but cannot come to any definite conclusion. To overcome set tasks ('pure cussedness' the Americans would say) is one reason (after once putting one's hand to the plough). To gain physical and mental strength, to raise and purify the mind in Nature's great school, are both potent reasons. But, above all, there is some mysterious influence pervading all true mountaineers—a mountain fever, a close kinship with Nature (call it what you will), a hidden impulse that grows on a man who has once felt what it is to taste the sweets of climbing and to enjoy the fascinations of the world above the snow-line.

My friend did not arrive, so I made my way over to Mr. Brodrick's Survey camp on the Murchison, walking through a thick mist, and steering across the Tasman by the aid of a compass—a distance of seven miles, or three hours' walking from camp to camp.

Here I found Cooper—Messrs. Wheeler & Son's photographic operator—who was down securing views of the district for a lecture which I was to deliver before the Australasian Association for the Advancement of Science.

It was our intention to make a two days' excursion up the Murchison Glacier with Cooper, but showery weather put a veto on our plans, and we were fain to be content with a short excursion to the Onslow Glacier, where some exposures were effected.

Leaving Mr. Brodrick's hospitable quarters on December 10, by the 12th we were again at the Hermitage.

CHAPTER X

ON SOME OF THE PHENOMENA OF GLACIERS, WITH SPECIAL REGARD TO THOSE OF NEW ZEALAND

The cause of glaciers—Formation and structure—Motion—Moraines: Lateral, medial, and terminal—' Surface' moraines—Crevasses—Moulins—Glacier cones—Glacier tables—Surface torrents—Avalanches—Cornices

IN a work of this nature it may not be out of place to briefly describe some of those interesting features and phenomena which accompany the world above the snow-line.

Here is a quotation from a recent review of Professor Heim's work [1] by a prominent member of the English Alpine Club :—

'Some thirty years ago a systematic *résumé* of all that was known up to that date about existing glaciers appeared in the work of Professor Albert Mousson, " Die Gletscher der Jetztzeit," since which, with perhaps the exception of Major Hüber's " Les Glaciers," no attempt has been made to collect into a focus the light which numerous able observers and theorists have subsequently thrown upon the question. The intricacy of the problem has, indeed, increased almost in proportion to our enlarged knowledge of its conditions ; and

---

[1] *Handbuch der Gletscherkunde*, von Dr. Albert Heim, Zürich (Stuttgart: Verlag von J. Engelhorn, 1885, 18 francs.)

in spite of the labours of a large and very distinguished body of investigators, not only do many important points remain matters of dispute, but the very materials for a complete solution are still wanting.'

CAUSE OF GLACIERS

The joint cause of glaciers is precipitation and cold. A low temperature alone can do nothing without moisture, and this fact quickly disposes of the popular notion that glaciers invariably exist in cold countries. Thibet, for instance, and also some parts of Arctic North America are destitute of ice streams, though eternal cold may be said to reign supreme in these parts.

Imagine for a moment the higher mountains clear of snow and ice, and then watch for the formation of a glacier. Snow falls and fills up all the valleys and gullies, avalanches descend from the higher parts, and a great accumulation gathers in all hollows. By constant repetition of snow-falls (always provided a greater quantity is deposited than can be melted by the sun's rays and by the natural warmth of the earth's crust) great pressure is put upon the lower portions by the superincumbent accumulation, and aided by the infiltration of water and refreezing (or 'regelation' as the correct term is), a large body of ice is formed which at once begins to move down the valleys containing it.

GLACIER ICE

Glacier ice is not like the solid blue ice on the surface of water, but consists of granules joined together by an intricate network of capillary water-filled fissures.

In exposed sections and upon the surface of the ice can be observed a 'veined' or 'banded' structure—veins of a denser blue colour alternating with those of a lighter shade containing air bubbles.

The cause of this peculiar structure has been the subject of much theorising amongst investigators, but hitherto I believe the greatest authorities consider that the explanation of the phenomenon is yet wanting.

GLACIER MOTION

The motion of glaciers is yet another bone of contention, but it is generally admitted that the cause of it is to be found mainly in gravitation, and is also partially accounted for by the strange property of 'viscosity' in what appears to the casual observer to be nothing more or less than a rigid solid.

Recently observations for ascertaining the rate of progress of the Tasman, Murchison, Hooker, and Mueller Glaciers have been made by the New Zealand Government Survey Department. Some of the results were embodied in a paper by Mr. J. H. Baker, the Chief Surveyor of the Provincial District of Canterbury, and will appear in the 'Transactions of the Australasian Association for the Advancement of Science' for 1891. At the late meeting of that body a committee was appointed to further these investigations, and a sum of 25*l.* voted for the aid of the same.

Before long, therefore, there will be put before the scientific public reliable measurements of the motion of several of the largest and least-known glaciers in temperate regions.

MORAINES

There is a remarkable feature of the glaciers of this country which stamps them as unique in one respect— I refer to the very extensive moraines. I write feelingly of this, for my acquaintance with them has been a very close one, and they have impressed me very deeply— in more ways than one.

The large glaciers of which I have written in this work are completely moraine-covered over their lower parts.

'SURFACE' MORAINES

Moraines may be divided into four sections: 'Lateral' moraines, fringing the sides of the glaciers, their outlying portions often being 'dead'—that is, at present unmoved by the action of the ice, and forming banks, as it were, for the ice stream to flow between; 'medial' moraines, which begin at the junction of two streams of ice and often continue for many miles to the terminal face; 'terminal' moraines, formed by the depositing of detritus at the melting point or end of the glacier; and, lastly, 'surface' moraines (so called by Professor Hutton of Christchurch, N.Z.), which are the combined accumulations of the first two divisions in the lower parts of the glacier.

It is these 'surface' moraines that are such a characteristic feature of the glaciers situate on the eastern side of the chain in New Zealand. Of those on the western side I am not able to speak with authority, never having visited them myself; but I understand that they do not carry such a large quantity of detritus as those of the eastern slopes.

This disparity remains to be accounted for and awaits an explanation. I have a theory of my own upon the subject, which, however, as yet I would not like to put too strongly forward.

On both sides of Mount Cook, on Mount De la Bêche (ten miles further along the chain), and on a peak just north of the Hochstetter Dome (ten miles still further north) I have observed enormous exposed sections of the rock strata, which in each case dip at a steep angle *from east to west*, presenting slab faces, not easily disturbed by the action of the frost, to the westward, but broken and fast denuding faces ('basset' faces, as they are geologically termed) to the eastward. I am hoping at some future time to further investigate this interesting subject.

As the western glaciers, however, must descend steeper valleys than the eastern, I make no doubt that their rate of progress will be eventually ascertained to be greater than that of the latter, and this would militate largely against an accumulation of moraine *upon the ice*.

THE SURFACE OF A GLACIER

All sorts of queer notions as to what the surface of a glacier is like exist. Indeed I have often heard people inquire if it would be possible *to skate upon it*!

Let us for a moment imagine ourselves at the head of the great Tasman Glacier, 8,600 feet above sea-level. All around us is snow, either freshly fallen or merging into *néré*. We begin to walk down, and at first, upon the steeper slopes, cross a few large crevasses and *bergschrunds* by means of snow bridges; then, as the incline becomes less steep, we walk for six miles or so upon a smooth surface of *névé*, or perchance knee-deep in fresh snow, and scarcely a crevasse exists. At the beginning of the great turn we gradually leave the *névé* and find ourselves upon hard, white ice, and soon transverse crevasses appear; these are a little further on cut by longitudinal crevasses forming the surface into huge squares, not flat on the top, but hummocky. A perfect network of crevasses cuts up the whole of the surface, but those parts on the outside of the curve are infinitely more disturbed than those on the inside, owing to the tension put upon them by the faster rate at which they have to move. After rounding the turn the glacier again consolidates and few crevasses appear, only the surface is covered with old wounds—if I may coin such a term—from the rents which have occurred at the turn, and presents a very undulating appearance. The little gullies are formed into watercourses and intersect the glacier in all directions. On our right, now, is the medial moraine formed by detritus from Mount De la Bêche, brought down partly by the Tasman and partly by the Rudolf Glaciers, and it stands up 100 feet or so above the surface of the clear ice on either side of it, owing to the protection from the sun's rays afforded by it to the ice beneath, so preventing 'ablation' or waste going on so quickly. We follow down for another four or five miles, and then cross this moraine (which has in the meantime joined that on the northern side of the Hochstetter Glacier) on to the Hochstetter on our right.

SURFACE TORRENTS AND MOULINS

We are now immediately below the great ice-fall, and the surface of the glacier presents an appearance not unlike the back of some enormous caterpillar wrinkled transversely by crevasses, which close up as we proceed downwards, and furrowed longitudinally by two large or main watercourses whose icy banks are in places 100 feet above their respective torrents. These two small rivers are fed from every direction by minor watercourses, and a mile or two further down discharge all their contents into crevasses and *moulins*, or water-shafts in the ice.

GLACIER TABLES AND CONES—THE ACTION OF WARMTH

The locality of the glacier on which we now are is very interesting, for Nature's mills are here seen at work day by day. Glacier tables—blocks of rock perched upon pedestals of ice formed by the protection from the action of the sun's warmth—are of frequent occurrence. Glacier cones—heaps of sand and small fragments of rock raised by a similar agency (after having been washed to one spot by water)—are in places all around us. Then, strange and contradictory as it may seem, we see thousands of holes, each with a stone at the bottom and filled with the bluest of blue water, formed also in the first place by the rays of the sun warming the stone and causing it to sink in the ice. It is well-known in physics that water at 39° Fahr. is at its heaviest, and as soon as the warm stone—the dark colour of the stone having absorbed more heat than the surrounding ice—begins to sink the warmer water follows it, whilst that in the neighbouring temperature of 32° Fahr. rises to the surface and becomes in its turn re-warmed, and so on. This peculiar current often bores the holes in the ice to a depth of many feet, and is only checked by a preponderance of cold. It is the larger stones, therefore, which rise upon the ice, and the smaller ones which sink.

'SURFACE' AND 'TERMINAL' MORAINES

We walk on down the ice stream, and soon the moraines on either hand close in upon us and we find ourselves on a mere wedge of ice, at the point of which we step on to the 'surface' moraine. Here the swearing begins, and it lasts right on to the terminal face four or five miles below, for it is one continual repetition of walking on loose and tumbling rocks, up one hillock, along a ridge, jumping from

Rock to rock with many a shock,

down another hillock, now and then starting a whole avalanche of many-sided and sharp-edged stones down a treacherous slope of ice, which we take for a surface deeply covered and sound of footing.

Skate on the surface of a glacier?

'Not much!' (as the Colonials say).

AVALANCHES

Very strange notions also exist amongst the uninitiated as to the nature of avalanches. The popular idea of an avalanche is derived from heartrending accounts of great sweepings away and annihilation of whole villages, and few of the general run of people seem to realise that in Alpine work almost any little descending mass of rock, snow, or ice is dignified by the name of avalanche. Snow avalanches are most frequent after fresh falls of snow followed immediately by warm weather, and after a little experience amongst the mountains one soon learns to detect their customary tracks. Ice avalanches are mainly caused through the overhanging portion of ice at the terminals of secondary glaciers—that is, glaciers which break off before descending to the valley or to the parent glacier below. The tracks of ice avalanches are almost invariably unmistakable and are swept night and day without cessation, and very frequently at regular intervals.

Rock avalanches are more treacherous, and one never knows when to expect them from above; generally in the early morning the frost holds the stones above in an icy grip, but as the sun melts the ice in the chinks the hold is released and a stone will descend into the *couloirs* or ditches which scarp the mountain side. If one happens to be below then it is a case of *sauve qui peut* and a rush for the nearest protection, for there is no saying how many tons, or indeed how many hundreds of tons, of loose rocks or stone may start in a wild and dusty rattle down the hillside.

But some snow avalanches almost crawl down the *couloirs*, and make a strange and ever-continued hissing as they move. These are composed of heavy and sodden snow, and begin after the sun has been up for some hours, continuing until nightfall. These are not so dangerous on a gentle slope, and one can often waddle or half glissade down in the midst of one with perfect safety, though they make one uncomfortably wet.

CORNICES

Cornices are a frequent source of danger to the mountaineer. They are formed by the snow drifting over one edge of a ridge and forming a hanging mass. It is needless to say that one soon learns to walk some feet away from the outer edge of a cornice, for after poking one's axe-handle through three feet of snow, and peeping through a blue hole down a precipice of perhaps 1,000 feet or so, it is not difficult to fancy what the result would be should the cornice break.

CHAPTER XI

CANOEING ON THE NEW ZEALAND RIVERS

The Waimakariri—The enormous rainfall—Descent of the Waitaki River—The Tasman branch—Lake Pukaki—Leaky canoes—The Pukaki Rapids—The Waitaki Gorge—Out on the plains again—Sixty miles' paddle to catch the train—Home once more

CANOEING on the New Zealand rivers is desperately exciting work. On the west coast of the South Island there is a canoe club, whose members build boats in watertight compartments specially suited for the rough journeys which they undertake. Some of these men are adepts at canoe-sailing, and think little of going out to sea in their cockleshells and even making long coastal journeys. The brothers Park have established quite a reputation by their adventurous journeyings. On one occasion they crossed the South Island with their canoes, towing up the Teramakau River, crossing a saddle of 1,700 feet at its head, descending the Hurunui and then coasting fifty miles down to Christchurch. On another occasion the crossing of Cook Straits was effected by them.

On the eastern side of the island not much canoeing has been done, with the exception of the navigation of two of the largest rivers (the Waimakariri and Waitaki) from their sources to the sea by Mr. Dixon and myself.

I well remember how universal was the outcry against our attempting to descend the Waimakariri in 1889, upon which occasion we conveyed the canoes up to the head waters in the Southern Alps, and came down ninety miles of rapids at a tremendous rate, going through the celebrated gorge fourteen miles in length. Dixon reached Christchurch in one day—a wonderful feat—but I was not able to accomplish more than half the distance, and took two days over it. This involved a descent of 2,550 feet in altitude from the starting-point.

In the following year the Waimakariri was again navigated by myself and three other kindred adventurous spirits, when a number of fine photographic views of the scenery in the gorge were secured.

The descent of the Waitaki River, however, promised some exciting work, in addition to giving a grand insight into the story of the ancient glacier formation—a subject of great geological interest.

The rainfall in the New Zealand mountain districts is enormously heavy, as much or more than 150 inches per annum being registered in some parts. The rivers consequently carry a phenomenal amount of water for their length, and the calculations as to their discharge give wonderful results. The Clutha River in Otago—the largest river of the South Island—discharges as much water per annum as the Nile! It seems a strange statement to make; but such is the fact, the calculations having been made by competent men.

The day following our return from Aorangi we left the Hermitage at 9 A.M., and by 1 P.M. had begun our exciting journey of 140 miles to the sea.

IN THE ICE-FALL OF THE ONSLOW GLACIER

The Tasman River takes its rise from the Tasman and Murchison Glaciers, and is soon joined by the Hooker, which drains the Hooker and Mueller Glaciers. Its course from Mount Cook to its delta at the head of Lake Pukaki is thirty miles in length, and the fall is considerable, the terminal face of the Tasman Glacier being 2,456 feet above sea-level, whilst the altitude of Lake Pukaki is 1,717 feet. The first mile or two of the journey was marked by several strong rapids, and we could not avoid shipping much water; and, added to this, we soon found that some old cracks in the canoes had opened out through exposure to the sun, although they had been carefully covered over with sacking during our absence in the mountains. This gave us some cause for anxiety, and the discomfort of paddling in boats which were half full of water soon made itself painfully apparent. Indeed, there is nothing more calculated to put a man out of temper with all the world and his surroundings, to goad him to strong language, and to give him an uncomfortable and miserable time generally, than to have to sit for hours in a boat that floats like an unmanageable log, to say nothing of the increase of danger to which he is consequently exposed in some parts of a river such as the Tasman, running, as it does, something approaching ten knots in many places.

I don't think Dixon and myself are likely to forget the tortures of the four hours which we passed through on reaching the lake. Here the cracks in my boat, which was decidedly the worse of the two, had to be jammed up with handkerchiefs, &c., before we dared to venture on a journey of eight or nine miles to the ferry at the other end of the lake, where is situated the exit of the Pukaki River.

As we scraped over the sandy shallows and pushed off into deep-green water, my heart sank within me at the idea of having to cross the lake in its present rough state (for a strong nor'-wester was blowing) in our frail canoes, which were not built in watertight compartments, and were quite unsuited for the work. Every ten minutes or so I would have to stop paddling and bale for dear life with the lid of the 'billy,' and the craft would immediately swing round broadside on to the seas, which seemed to do their best to upset her.

At first we kept edging away for the southern shore, and about half-way down the lake succeeded in getting within reasonable swimming distance, which, to a certain extent, we retained for a short time.

In the distance we could make out the island close to the ferry, with some trees on it, and from our direction there appeared to be but three. My thoughts at once flew back to the island on the Lake of Geneva, which Byron has immortalised in his 'Prisoner of Chillon,' and on which poor Bonnivard would gaze with sadness and yearning for freedom and life.

> And then there was a little isle,
> Which in my very face did smile,
>   The only one in view.
> A small green isle, it seemed no more,
> Scarce broader than my dungeon floor;
> But in it there were three tall trees,
> And o'er it blew the mountain breeze.
> And by it there were waters flowing,
> And on it there were young flowers growing
>   Of gentle breath and hue.

I made sure my hair would be grey, like poor Bonnivard's, before this lake was crossed; but soon the wind dropped, and we paddled ashore at 9 P.M. close to the hotel and called for brandy and water hot, and seldom was the indulgence more justified.

At Pukaki Ferry we enjoyed a well-earned night's rest, and on Sunday morning we effected repairs to the leaky canoes, in which operation we received much valuable advice and assistance from Mr. John Gibb, artist, who was spending a few days in sketching at this point. By 1 P.M. we were on board again and looking forward to reaching Rugged Ridges—Mr. W. G. Rutherfurd's station on the southern bank of the Waitaki—before nightfall. But we little knew what was ahead of us.

A survey of the river from an eminence of the old moraine through which it has formed a channel, revealed, as far as the bends of the stream could be followed, a rushing, seething mass of foam-covered water, with numberless blocks of rock barring the clear passage of the current, and though we shot the first two rapids below the exit from the lake it took us until seven o'clock in the evening to navigate six miles of the river's course.

It is not easy to describe the wild course of the river in its descent through the enormous ancient moraine deposits, some of which might almost be classed as mountains, and must rear their tops to a height of 1,000 feet above the level of the river. Such an immense body of rushing water, receiving, as it does, the whole of the drainage of the Southern Alps, from the head of the Mueller Glacier to that of the Murchison, necessarily creates great havoc amongst the glacial and fluviatile deposits through which it descends, and, as a matter of course, all the smaller stones are hurried and rolled along to form shingle on the river-beds further down, leaving the larger ones, which alone can stand against the force of the flood. The natural consequence is a stream of the most broken and impetuous character, a stream whose rushing, roaring, and foaming drowns all sounds contiguous to it; rapid after rapid of seemingly tempest-tossed and crested billows, of whirlpools and eddies, of backwaters and heavings into surface currents, and never a still pool to be found anywhere.

Imagine, then, the troubles of two canoeists in navigating this stretch of water. No canoe or boat in the world would have the slightest chance of going through, out in the current, without being smashed into matchwood and its occupants infallibly drowned, for swimming would avail a man nothing in such a place.

All we could do, then, was to keep close to the bank and let our frail boats down by the tow-lines amongst the rocks in the comparatively shallow water. Now shoving them off into a fair stretch and hauling them up short in time to avoid contact with some ugly rock in front, then scrambling along ourselves and coiling our lines as we advanced, clambering over water-worn and slippery rocks, tearing our way through the Wild Irishman scrub, or wading a few steps middle-deep in the turbid water to the points where we had brought our respective canoes up. Then repeating the same performance over again and again, bruising our legs against rocks, slipping down amid the slimy stones, scratching the skin off and receiving numerous thorns from the scrub, wishing we had never been born, lamenting the hardships of our lot, anathematising canoes, ropes, paddles, river, rocks, scrub, and everything in creation.

No, that seven miles journey was *not* all that could be desired; but having put our hands to the plough, we both made up our minds that we would go through with the undertaking, even if we had to repeat the same performance down to the sea every day for a week, and the worse the river got the more pig-headed we became. We had beaten Mount Cook, and we meant also to gain a victory over the Pukaki and Waitaki, if it cost us our life-blood. At some places where a number of large rocks were congregated close to the river's bank we would be compelled to take the boats out, and shouldering them, climb round the rocks on shore, and launch them afresh in better water below.

At one time, Dixon, who was leading, accidentally dropped his paddle, which was whisked away by the current in a trice. He made a great effort to recover it, and plunged in up to his armpits in the turbulent water, but failed to reach the truant paddle. Seeing his difficulty I pushed my boat out to him, and he seized my paddle and, jumping into the canoe, gave chase to the one he had lost. I ran along the bank, but could not keep near him; and in fear and trembling I watched him nearing a horrible fall amongst some sharp teeth-like rocks. I thought his last moment had come, but just before reaching the danger he overtook the lost paddle, which he grasped with one hand, and, jumping out of my canoe, held the tow-rope and brought the boat up within a few feet of the fall. The whole affair was the work of a few moments, and was a wonderful exhibition of smartness and presence of mind.

By 7 o'clock we began to think that we had had about enough for the day, and, putting the boats ashore, we walked back, over the old moraine and along the rabbit-fence (which, by the way, I hear is doing its work splendidly), to the Pukaki Ferry for the night.

By 7 A.M. next morning we were again with the canoes, and once more performing gymnastic feats along the rocky bank. But our reward was now near at hand, for after an hour or so we got on board and sneaked down the quieter sides of one or two pools. The moraine deposits gave way to those of fluviatile origin, and the size of the stones in the river-bed decreased rapidly; consequently we soon began shooting the rapids again and were making grand headway. The country on either hand opened out; from our left came in the Tekapo River, and soon after, as we sped on under Ben More, on our right the Ohau. Now we were in the Waitaki, which is formed by the junction of these three rivers. 'Waitaki,' or 'Waitangi,' means 'Crying water.'

The hydrographic area of the Waitaki Basin is 4,914 square miles, more than three times as great as that of the Rakaia or Waimakariri, and it drains most of the principal eastern slopes of the Southern Alps.

The eastern source of the river drains the Godley and Classen Glaciers with their numerous tributaries, forms the Godley River, and flows into Lake Tekapo (some fifteen miles in length); it issues from the southern end of the Lake and curves a channel for itself through the ancient moraine, when it becomes known by the name of the Tekapo River, which, flowing for a distance of about twenty-five miles, joins the Pukaki; all these, with the addition of the Ohau, the junction of which is a few miles further down, form the Waitaki River. The Hopkins and Dobson Rivers drain that part of the Alps immediately south-west of Mount Sefton, and flow into Lake Ohau. The stream issuing from thence, under the name of the Ohau River, runs for a course of thirteen miles, and joins the Pukaki and Tekapo as before mentioned.

After the union of these three systems of drainage the course of the river runs through a wider bed for about five or six miles before entering a gorge some ten miles in length. Down this fine stretch of water we now enjoyed a delightful paddle, and soon we sighted Black Forest sheep station, with its rows of green willow trees, on our left.

Here various kinds of river birds lent an aspect of life and gaiety to the scene—gulls, terns, paradise and grey duck, teal, dotterel, stilt, and red-bill soared over us, or rose in startled dismay as we shot by.

We had left the snows behind us and were fast being closed in by the foot-hills. We neared the gorge at 11 A.M. and paddled ashore on the Otago side and boiled the 'billy' for lunch.

It seemed a delightfully quiet hour after all we had been through; we sat and smoked in happiness and watching the rabbits skipping about amongst the bracken. We were certain, if only by that, that we were in Otago, where rabbits are the monarchs of all they survey.

The Mackenzie country hands had told us that we should find the gorge *a little rough*, so we knew we were in for it presently; yet for a couple of miles we found the river good going, though some ominous spurs of bed rock now and then entering the current—the first bed rock we had met with since leaving Mount Cook—foretold what we were coming to.

After going round a few ugly corners the white water became more frequent, until suddenly we were brought up by an awkward rapid into which we dared not venture.

A survey from the cliffs, sixty feet above the stream, disclosed a tongue or groyn of rocks running out into the stream in an oblique direction from the Otago side, and shooting the main body of the current on to the rocks opposite. A long stretch of straight water followed, but the whole stream was confined in rocky banks so close together that one might throw a biscuit across, and the pace of the current was something terrific. For half an hour we considered the situation, finally determining to shoot the rapid. There was really only about eight or ten feet of safe water close to the point of the groyn of rocks, and this was right in the body of the current. On either hand were eddies and whirlpools of the most formidable character, which, in the event of our making a bad shot, might swirl us among the rocks on one side or the other, and had such been the case we trembled to think what would have been our fate. However, at it we went, Dixon as usual leading, with a head as cool as a

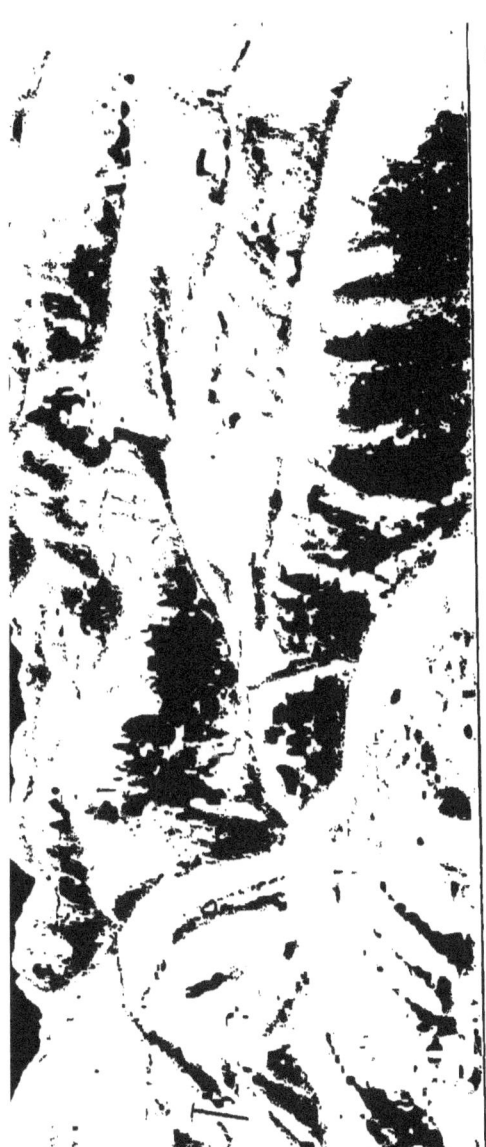

THE SURFACE OF A GLACIER

[*Wheeler & Son, Photo.*

cucumber, and I following, like a spaniel after his master. One wild rush, a few strokes of the paddle, a mad tossing about in a sheet of crested foam, half-a-dozen bucketfuls of water on board, and we were through, breathing again as we tore down the hurrying, but straight and safe, current below.

Though we met with no greater obstacles to canoeing than this rapid in the gorge, such performances were several times repeated, and we had to land now and again to survey the course ahead.

To describe the mad plunging of the river through the gorge is not an easy matter. Here and there, perhaps, a long even stretch is met with, but for the most part the river makes a succession of bends bounded by rocky cliffs on either hand, now and then masses of rock crop up through the water, against which the stream is banked up by the force of its mad career to a height of ten or twelve feet; immediately under the sides of the rock there are vicious-looking heavings, eddies, and whirlpools, which, if one chances to get into them, twist the boat about like a feather when blown upon the water's surface. A black swan and three cygnets kept ahead of us for the last six miles of the gorge, but as we entered with relieved feelings upon the more open country, they eluded our further pursuit in a backwater. Another few miles and we reached our destination for the night—Mr. W. G. Rutherfurd's station, Rugged Ridges—where a warm and hospitable welcome made us feel that once more we were in the regions of civilisation.

Leaving next morning at 4.30, we gave ourselves eleven hours to catch the train for Christchurch, at

Waitaki, a distance by water of sixty miles. Four hours saw us in Duntroon (thirty miles), where we astonished the natives in disgracefully tattered boating attire, and indulged in that from which we had long been estranged—'a long shandy'—and by 9.15 we were off again at eight miles an hour, shooting down the most beautifully safe and rippling rapids, scaring ducks, plover, gull, stilt, swan, and all manner of wild fowl; now and then startling a mob of horses or cattle from their peaceful browsing, or astonishing some slow-going shepherd or cowboy as they stared open-mouthed at such an uncommon sight as two madmen in cockle-shells of canoes rushing down their boatless river, until we put the final touch to the whole enterprise by carrying our boats up to the station at Waitaki South (to the amazement of four railway navvies), at 1 P.M., having averaged eight miles an hour for sixty miles, allowing for one hour stoppages.

The distances by water, allowing for sinuosities in the course of the rivers from Aorangi to the sea, may be roughly summarised as follows:—From the end of the Mount Cook Range to Pukaki Ferry, thirty-four miles; from the Ferry to Rugged Ridges, thirty-eight miles; and from thence to the railway bridge near the sea at Waitaki, sixty miles; a total distance of 132 miles.

If it were not for the Pukaki Rapids the trip might be comfortably accomplished in three days, and at a stretch could be done in two; but the way to enjoy it would be to travel in a good staunch canoe, with watertight compartments and such accessories as the west coast canoeists are in the habit of using, and spend a week over the journey.

L'ENVOI

This little book has but told the story of the ramblings and adventures of a lover of Nature. I fear that I have signally failed to do justice to her features, or to convey any adequate idea of her mystic influence. Would that I could impart that which I can feel.

Should it fall into the hands of Swiss climbers it may serve to show that the brotherhood of the mountains extends even to out-of-the-way New Zealand, and that in that country, as well as in the Old World, the ineffable glories of the mountains have power to charm and to captivate the hearts of men.

APPENDIX

It has been suggested to me that this work would be more complete if it contained at least a short record of Alpine expeditions undertaken by parties (other than those organised by the writer) to the glacier regions which have been under notice. The suggestion is one which the writer accepts with much pleasure.

The first recorded expedition to the Mount Cook district, as far as I am able to gather, is that of the late Sir Julius von Haast (then Dr. von Haast), the narrative of which may be found in his interesting and learned work 'The Geology of Canterbury and Westland,' published by the 'Times' office of Christchurch, now unfortunately out of print, and difficult of access to the majority.

His work was necessarily more that of exploration than of climbing, and although later surveys have corrected and modified many of his estimates of the sizes of glaciers and heights of mountains, it must be remembered that in the days when he visited the locality (in 1862 and 1870) the difficulties of travelling and of securing supplies were much greater than at the present time, and the work of exploration consequently much more difficult.

Of Alpine work (carried on in the sense of the word as understood by Alpine climbers) he did not effect much, his energies being chiefly confined to geological, botanical, and zoological observations whilst he was engaged in a geological survey of the province of Canterbury.

His excursions on the glaciers appear to have been confined to a short trip up the Tasman, probably to some six miles or so from the terminal face, and a short exploration of the lower portions of the Mueller and Hooker Glaciers.

His literary contributions are of greater value to science than to the domain of Alpine record; but naturally they are of the deepest interest to the latter class of literature, inasmuch as they tell the tale of the opening out of fresh Alpine fields which are destined to become—indeed they are now fast becoming—areas of great mountaineering importance.

Though Von Haast was perhaps the first man of science or literature to visit these great glaciers, yet their existence was well known to a few run-holders and early settlers who had penetrated even thus far into the mountains in the 'early days' of New Zealand.

It is to Mr. Edward Percy Sealy of Timaru, however, that we owe the first close acquaintance of the Mueller, Hooker, and Tasman Glaciers. Mr. Sealy was a surveyor by profession and a photographer of no mean ability, and to his energy and perseverance we are indebted for results which furnished Dr. von Haast with material for constructing his map of this part of our Alps.

Upon visiting the glaciers at the present time, and being impressed with the difficulties of transit, one cannot but be filled with admiration for the man who achieved such splendid results in photography, burdened as he was with all the necessary and cumbersome paraphernalia pertaining to the old wet-plate system then in vogue.

Mr. Sealy traversed nearly the whole length of the Mueller Glacier in 1867, and in 1869 pushed his way up the Hooker as far as the tributary Empress Glacier, and up the Tasman as far as the great turn at Mount De la Bêche.

To Mrs. Leonard Harper, of Ilam, belongs the honour of being the first lady to cross to the Aorangi side of the Tasman River.

On this occasion (in March 1873) the party consisted of Mr. and Mrs. Leonard Harper, of Christchurch, Messrs. G. Dennistoun, G. Parker, Melville Gray, Wright, C. Smith, and Flint. They camped at Governor's Bush, close to where the Hermitage now stands, and went on to the Mueller Glacier and to the terminal face of the Tasman. Mr. and Mrs. Harper returned across the Tasman River, leaving the rest of the party to attempt the passage to the west coast by the Hooker Saddle, at the head of the glacier of the same name. In this, as may be easily conceived—considering that the members of the party were inexperienced and not properly equipped for such an expedition—the party was unsuccessful, only reaching a point just above where the clear ice merges into the moraine, and where the crevasses began to appear formidable.

For many years after this the glaciers were not traversed to any extent save by camping-out parties, who contented themselves with short excursions about the terminal faces, until, in 1882, a fresh interest was awakened in their existence by the visit of the Rev. W. S. Green with Herr Emil Boss, of Grindelwald, and Ulrich Kaufmann as guide. His advent was indeed an awakening, and the apathy of the Colonials regarding the scenic marvels of their own country was somewhat aroused. The sensation caused by his memorable ascent of Aorangi, after repeated struggles with flooded rivers and all those hindrances which seem to fall inevitably to the lot of men who first open out a new district, has become quite an event of history in the annals of the colony.

Full particulars of Mr. Green's doings will be found in his admirable book, 'The High Alps of New Zealand,' published by Macmillan & Co.

To Mr. Green undoubtedly belongs the honour of having first introduced into New Zealand the proper system of Alpine climbing, and he will ever be looked back to as the father of the noble sport in the colony.

Then, in 1883, followed the visit of Dr. R. von Lendenfeld, a mountaineer and scientific man of great attainments. He was accompanied by his plucky wife, and, aided by porters procured in the colony, during a stay of nineteen days on the Tasman Glacier completed a survey of the same, and finished up his work by ascending the Hochstetter Dome, whose higher and easternmost summit he attained in an expedition extending over a period of twenty-seven hours from his last camp under the Malte Brun range, accompanied by his wife and one porter.

Full particulars of his work were made public in Petermann's

'Mitteilungen,'[1] and a short English notice of the same may be found in the 'Alpine Journal,' vol. xii. page 163.

Shortly after this the Hermitage Company, Limited, was formed, and the Hermitage Hotel erected near the terminal face of the Mueller Glacier. This first Alpine hotel of New Zealand was not built without many serious difficulties, and the ultimate success of the undertaking speaks volumes for the perseverance of the enthusiastic manager, Mr. F. F. C. Huddleston. This gentleman has made various excursions on the Mueller and Hooker Glaciers since the building of the Hermitage, and possesses an intimate knowledge of the Alpine district around the hotel. He has, with a party of two others, penetrated, I understand, as far as the junction of the Empress Glacier on the Hooker, and has since effected the passage of the Ball Pass from the Tasman to the Hooker Glaciers.

In 1886 the author began his visits to the districts with properly equipped Alpine parties, the results of which expeditions have been given in the foregoing pages.

In 1889 the Government surveys were extended to the Mueller and Hooker Glaciers, under Mr. Brodrick, a gentleman whose capability and never-failing pluck in carrying out his work in such rough country is only equalled by his modesty concerning his Alpine achievements, which are necessarily incidental to his profession in the district.

Those of my readers who are acquainted with survey and topographical work amongst the Alps will appreciate the results of but two seasons' work in the map of the four great glaciers appended to this book. Climbers will be interested to know that amongst difficult points attained by Mr. Brodrick are the saddles at the head of the Mueller Glacier, that connecting the Murchison and Classen Glaciers, the lower summit of the Hochstetter Dome, and a peak of 8,015 feet on the Liebig Range.

In 1890, Mr. Malcolm Ross, of Dunedin, a gentleman who has done much travelling and some exploring in the Southern Lakes district, and had tried his 'prentice hand upon Mount Earnslaw, visited the Tasman Glacier in company with

[1] *Ergänzungsheft*, No. 75. Dr. R. von Lendenfeld, *Der Tasman-Gletscher und seine Umgebung*.

his wife. Bad weather frustrated his attempts at mountaineering, with the exception of an ascent of a peak of about 7,000 feet on the Mount Cook Range, and a partial ascent of Mount Sealy. He traversed the Tasman Glacier to a point some miles beyond the junction of the Hochstetter Glacier.

In December 1890 Messrs. A. P. Harper, R. Blakiston, and Beadel made an excursion to the Tasman Glacier, but bad weather kept them prisoners at camp nearly all the time of their stay. Messrs. Harper and Blakiston, after retreating from the Tasman, succeeded in reaching for the first time the saddle at the head of the Hooker Glacier (8,580 feet), after a trying expedition. This had been attempted several times before, but owing to numberless crevasses was found to be unattainable. Being early in the season and after a considerable snow-fall, however, the party in question found the crevasses mostly covered, and they were aided, moreover, by Mr. Harper's skill and knowledge of Alpine work.

Again, in January 1891, Messrs. Harper and Johnson visited the Tasman Glacier, and besides attaining a high saddle (about 7,500 feet) in the Malte Brun Range and making a nearly complete ascent of Mount Sealy, secured a fine collection of photographs.

Such, in brief, is a history of what Alpine work has been accomplished amongst the Southern Alps of New Zealand. Other glacier excursions, it is true, there have been, but they are few, and with the exception of the expeditions of Mr. Sealy and the Government Survey to the Godley and Classen Glaciers farther north, and of a few attempts to climb Mount Earnslaw in the Southern Lake district, are not worthy of much note as Alpine expeditions, undertaken in the orthodox manner with axe and rope.

As these lines are being penned the New Zealand Alpine Club is in process of formation, and the writer hears with pleasure of the probabilities of success which are likely to attend the efforts of the promoters of the club.

Letters of advice and encouragement from prominent members of the English Alpine Club have at various times come to

hand, and the friendly interest of mountain explorers of the
early days of the colony gives promise of an auspicious birth
to one of those bodies of enthusiasts whose aims may not be mercenary and self-seeking, but whose operations may contribute
their little unit to the art, literature, and scientific observation
of the times.

Who can say what the future may bring forth in the matter
of Alpine climbing in New Zealand? There is an immense
field—magnificent glaciers and noble peaks without number, as
yet practically untouched.

One can already see visions of parties of enthusiasts threading their way amongst intricate ice-falls, cutting steps up hard
ice slopes, conquering by persistent effort splendid rock peaks,
drinking in the glories of a new and fascinating world. Not
climbing from a gymnast's point of view, but climbing because—why? They cannot tell you why; but because they feel
and know the physical and spiritual benefits of a closer contact
with Nature, with an Omnipotent and Ever-guiding Hand,
which rules all things and creates a heaven even upon earth.

A SHORT GLOSSARY OF TECHNICAL ALPINE TERMS.

Arête.—A ridge either of rock, ice, or snow, or combinations of all three.

Bergschrund.—The crevasse or deep moat almost invariably found between the sides and upper portions of a glacier or ice slope and the rocks above, or the permanent clinging ice above, as the case may be. Of late the meaning of the term has become extended, and almost any crevasse in the upper parts of a glacier with one lip higher than the other comes under the designation.

Col.—Saddle, or dip in a ridge.

Cornice.—The overhanging edge of an *arête* caused by drifting snow.

Couloir.—A ditch or deep gully in the mountain-side; in the upper regions being usually floored with ice and swept by avalanches.

Crevasse.—The rent caused by fracture of the ice under tension.

Gendarme, or *rock tower.*—A mass of rock on the crest of an *arête.*

Moraine.—The accumulation of detritus which has fallen from the mountains on to the ice and is carried down upon it.

Névé, or *firn.*—Snow in a transition stage between snow and ice. The large fields of this feeding a glacier are spoken of as the *névés* of the glacier.

Séracs.—Blocks of ice broken into polyhedral masses (mostly cubic) by the body of the ice being crevassed in various lines of fracture. So called from the resemblance the blocks bear to a certain kind of cheese.

Shale slips and *shingle and boulder fans* are of very common occurrence in the New Zealand mountains and are caused by the discharge of detritus down *couloirs*, from which when emerging it spreads out into fan-shaped slopes.

RS

by
30
rom Surveys

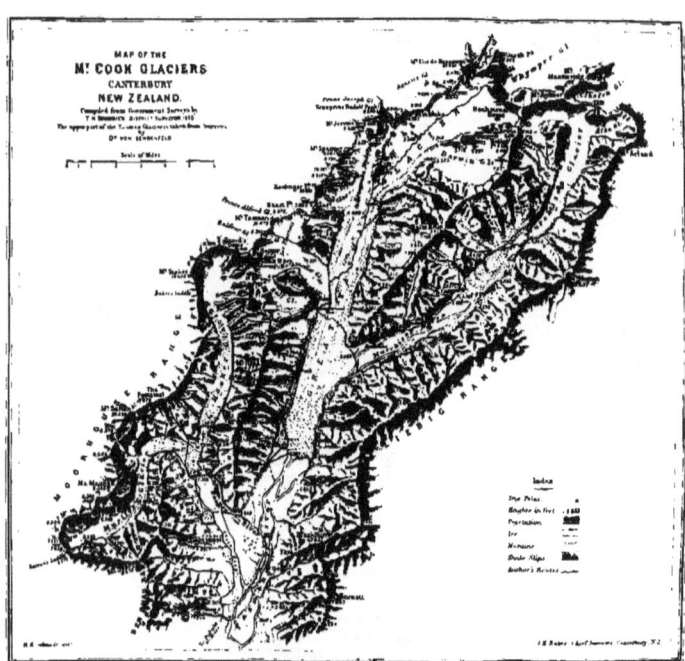

A CATALOGUE OF WORKS

IN

GENERAL LITERATURE

PUBLISHED BY

MESSRS. LONGMANS, GREEN, & CO.

39 PATERNOSTER ROW, LONDON, E.C.

MESSRS. LONGMANS, GREEN, & CO.

Issue the undermentioned Lists of their Publications, which may be had post free on application:—

1. MONTHLY LIST OF NEW WORKS AND NEW EDITIONS.
2. QUARTERLY LIST OF ANNOUNCEMENTS AND NEW WORKS.
3. NOTES ON BOOKS; BEING AN ANALYSIS OF THE WORKS PUBLISHED DURING EACH QUARTER.
4. CATALOGUE OF SCIENTIFIC WORKS.
5. CATALOGUE OF MEDICAL AND SURGICAL WORKS.
6. CATALOGUE OF SCHOOL BOOKS AND EDUCATIONAL WORKS.
7. CATALOGUE OF BOOKS FOR ELEMENTARY SCHOOLS AND PUPIL TEACHERS.
8. CATALOGUE OF THEOLOGICAL WORKS BY DIVINES AND MEMBERS OF THE CHURCH OF ENGLAND.
9. CATALOGUE OF WORKS IN GENERAL LITERATURE.

ABBEY and OVERTON.—**The English Church in the Eighteenth Century.** By CHARLES J. ABBEY and JOHN H. OVERTON. Cr. 8vo. 7s. 6d.

ABBOTT.—**Hellenica.** A Collection of Essays on Greek Poetry, Philosophy, History, and Religion. Edited by EVELYN ABBOTT, M.A., LL.D., Fellow and Tutor of Balliol College, Oxford. 8vo. 16s.

ABBOTT (*Evelyn, M.A., LL.D.*)—WORKS BY.

A Skeleton Outline of Greek History. Chronologically Arranged. Crown 8vo. 2s. 6d.

A History of Greece. In Two Parts.
Part I.—From the Earliest Times to the Ionian Revolt. Crown 8vo. 10s. 6d.
Part II. Vol. I.—500-445 B.C. [*In the Press.*
Vol. II.—[*In Preparation*].

ACLAND and RANSOME.—**A Handbook in Outline of the Political History of England to 1890.** Chronologically Arranged. By A. H. DYKE ACLAND, M.P., and CYRIL RANSOME, M.A. Crown 8vo. 6s.

ACTON.—**Modern Cookery.** By ELIZA ACTON. With 150 Woodcuts. Fcp. 8vo. 4s. 6d.

A. K. H. B.—*THE ESSAYS AND CONTRIBUTIONS OF.* Crown 8vo.

Autumn Holidays of a Country Parson. 3s. 6d.
Changed Aspects of Unchanged Truths. 3s. 6d.
Commonplace Philosopher. 3s. 6d.
Counsel and Comfort from a City Pulpit. 3s. 6d.
Critical Essays of a Country Parson. 3s. 6d.

[*Continued on next page.*

A. K. H. B.—THE ESSAYS AND CON-TRIBUTIONS OF—continued.

  East Coast Days and Memories. 3s. 6d.
  Graver Thoughts of a Country Parson. Three Series. 3s. 6d. each.
  Landscapes, Churches, and Moralities. 3s. 6d.
  Leisure Hours in Town. 3s. 6d.
  Lessons of Middle Age. 3s. 6d.
  Our Little Life. Two Series. 3s. 6d. each.
  Our Homely Comedy and Tragedy. 3s. 6d.
  Present Day Thoughts. 3s. 6d.
  Recreations of a Country Parson. Three Series. 3s. 6d. each.
  Seaside Musings. 3s. 6d.
  Sunday Afternoons in the Parish Church of a Scottish University City. 3s. 6d.
  'To Meet the Day' through the Christian year; being a Text of Scripture, with an Original Meditation and a Short Selection in Verse for Every Day. 4s. 6d.

American Whist, Illustrated: containing the Laws and Principles of the Game, the Analysis of the New Play and American Leads, and a Series of Hands in Diagram, and combining Whist Universal and American Whist. By G. W. P. Fcp. 8vo. 6s. 6d.

*AMOS.—***A Primer of the English Constitution and Government.** By SHELDON AMOS. Crown 8vo. 6s.

Annual Register (The). A Review of Public Events at Home and Abroad, for the year 1890. 8vo. 18s.
  *** Volumes of the 'Annual Register' for the years 1863-1889 can still be had.

ANSTEY (F.)—WORKS BY.

  The Black Poodle, and other Stories. Crown 8vo. 2s. bds.; 2s. 6d. cl.
  Voces Populi. Reprinted from *Punch.* With 20 Illustrations by J. BERNARD PARTRIDGE. Fcp. 4to. 5s.

ARISTOTLE.—THE WORKS OF.

  The Politics: G. Bekker's Greek Text of Books I. III. IV. (VII.), with an English Translation by W. E. BOLLAND, M.A.; and short Introductory Essays by A. LANG, M.A. Crown 8vo. 7s. 6d.
  The Politics: Introductory Essays. By ANDREW LANG. (From Bolland and Lang's 'Politics'.) Crown 8vo. 2s. 6d.
  The Ethics: Greek Text, Illustrated with Essays and Notes. By Sir ALEXANDER GRANT, Bart., M.A., LL.D. 2 vols. 8vo. 32s.
  The Nicomachean Ethics: Newly Translated into English. By ROBERT WILLIAMS, Barrister-at-Law. Crown 8vo. 7s. 6d.

ARMSTRONG (G. F. Savage-)—WORKS BY.

  Poems: Lyrical and Dramatic. Fcp. 8vo. 6s.
  King Saul. (The Tragedy of Israel, Part I.) Fcp. 8vo. 5s.
  King David. (The Tragedy of Israel, Part II.) Fcp. 8vo. 6s.
  King Solomon. (The Tragedy of Israel, Part III.) Fcp. 8vo. 6s.
  Ugone: A Tragedy. Fcp. 8vo. 6s.
  A Garland from Greece; Poems. Fcp. 8vo. 9s.
  Stories of Wicklow; Poems. Fcp. 8vo. 9s.
  Mephistopheles in Broadcloth: a Satire. Fcp. 8vo. 4s.
  The Life and Letters of Edmund J. Armstrong. Fcp. 8vo. 7s. 6d.

ARMSTRONG (E. J.)—WORKS BY.
  Poetical Works. Fcp. 8vo. 5s.
  Essays and Sketches. Fcp. 8vo. 5s.

ARNOLD (Sir Edwin, K.C.I.E.)—WORKS BY.
  The Light of the World; or, the Great Consummation. A Poem. Crown 8vo. 7s. 6d. net.
  Seas and Lands. With numerous Illustrations. 8vo.

ARNOLD (Dr. T.)—WORKS BY.
  Introductory Lectures on Modern History. 8vo. 7s. 6d.
  Sermons Preached mostly in the Chapel of Rugby School. 6 vols. Cr. 8vo. 30s., or separately 5s. ea.
  Miscellaneous Works. 8vo. 7s. 6d.

*ASHLEY.—***English Economic History and Theory.** By W. J. ASHLEY, M.A. Part I. The Middle Ages. 5s.

Atelier (The) du Lys; or, An Art Student in the Reign of Terror. By the Author of 'Mademoiselle Mori'. Crown 8vo. 2s. 6d.

BY THE SAME AUTHOR.
  Mademoiselle Mori: a Tale of Modern Rome. Crown 8vo. 2s. 6d.
  That Child. Illustrated by GORDON BROWNE. Crown 8vo. 2s. 6d.

[Continued on next page.

Atelier (The) du Lys — *WORKS BY THE AUTHOR OF — continued.*

Under a Cloud. Cr. 8vo. 2s. 6d.

The Fiddler of Lugau. With Illustrations by W. RALSTON. Crown 8vo. 2s. 6d.

A Child of the Revolution. With Illustrations by C. J. STANILAND. Crown 8vo. 2s. 6d.

Hester's Venture: a Novel. Cr. 8vo. 2s. 6d.

In the Olden Time: a Tale of the Peasant War in Germany. Cr. 8vo. 2s. 6d.

BACON.— THE WORKS AND LIFE OF.

Complete Works. Edited by R. L. ELLIS, J. SPEDDING, and D. D. HEATH. 7 vols. 8vo. £3 13s. 6d.

Letters and Life, including all his Occasional Works. Edited by J. SPEDDING. 7 vols. 8vo. £4 4s.

The Essays; with Annotations. By RICHARD WHATELY, D.D., 8vo. 10s. 6d.

The Essays; with Introduction, Notes, and Index. By E. A. ABBOTT, D.D. 2 vols. fcp. 8vo. price 6s. Text and Index only, without Introduction and Notes, in 1 vol. Fcp. 8vo. 2s. 6d.

The BADMINTON LIBRARY, Edited by the DUKE OF BEAUFORT, K.G., assisted by ALFRED E. T. WATSON.

Hunting. By the DUKE OF BEAUFORT, K.G., and MOWBRAY MORRIS. With 53 Illus. by J. Sturgess, J. Charlton, and A. M. Biddulph. Cr. 8vo. 10s. 6d.

Fishing. By H. CHOLMONDELEY-PENNELL.
Vol. I. Salmon, Trout, and Grayling. With 158 Illustrations. Cr. 8vo. 10s. 6d.
Vol. II. Pike and other Coarse Fish. With 132 Illustrations. Cr. 8vo. 10s. 6d.

Racing and Steeplechasing. By the EARL OF SUFFOLK AND BERKSHIRE, W. G. CRAVEN, &c. With 56 Illustrations by J. Sturgess. Cr. 8vo. 10s. 6d.

Shooting. By LORD WALSINGHAM and Sir RALPH PAYNE-GALLWEY, Bart.
Vol. I. Field and Covert. With 105 Illustrations. Cr. 8vo. 10s. 6d.
Vol. II. Moor and Marsh. With 65 Illustrations. Cr. 8vo. 10s. 6d.

Cycling. By VISCOUNT BURY (Earl of Albemarle), K.C.M.G., and G. LACY HILLIER. With 19 Plates and 70 Woodcuts, &c., by Viscount Bury, Joseph Pennell, &c. Crown 8vo. 10s. 6d.

The BADMINTON LIBRARY — *continued.*

Athletics and Football. By MONTAGUE SHEARMAN. With 6 full-page Illustrations and 45 Woodcuts, &c., by Stanley Berkeley, and from Photographs by G. Mitchell. Crown 8vo. 10s. 6d.

Boating. By W. B. WOODGATE. With 10 full-page Illustrations and 39 woodcuts, &c., in the Text. Cr. 8vo. 10s. 6d.

Cricket. By A. G. STEEL and the Hon. R. H. LYTTELTON. With 11 full-page Illustrations and 52 Woodcuts, &c., in the Text, by Lucien Davis. Cr. 8vo. 10s. 6d.

Driving. By the DUKE OF BEAUFORT. With 11 Plates and 54 Woodcuts, &c., by J. Sturgess and G. D. Giles. Crown 8vo. 10s. 6d.

Fencing, Boxing, and Wrestling. By WALTER H. POLLOCK, F. C. GROVE, C. PREVOST, E. B. MICHELL, and WALTER ARMSTRONG. With 18 Plates and 24 Woodcuts, &c. Crown 8vo. 10s. 6d.

Golf. By HORACE HUTCHINSON, the Rt. Hon. A. J. BALFOUR, M.P., ANDREW LANG, Sir W. G. SIMPSON, Bart., &c. With 19 Plates and 69 Woodcuts, &c. Crown 8vo. 10s. 6d.

Tennis, Lawn Tennis, Rackets, and Fives. By J. M. and C. G. HEATHCOTE, E. O. PLEYDELL-BOUVERIE, and A. C. AINGER. With 12 Plates and 67 Woodcuts, &c. Crown 8vo. 10s. 6d.

Riding and Polo. By Captain ROBERT WEIR, Riding Master, R.H.G., and J. MORAY BROWN. With Contributions by the Duke of Beaufort, K.G., the Earl of Suffolk and Berkshire, the Earl of Onslow, E. L. Anderson, and Alfred E. T. Watson. With 18 Plates and 41 Woodcuts, &c. Crown 8vo. 10s. 6d.

BAGEHOT (Walter).— WORKS BY.

Biographical Studies. 8vo. 12s.

Economic Studies. 8vo. 10s. 6d.

Literary Studies. 2 vols. 8vo. 28s.

The Postulates of English Political Economy. Cr. 8vo. 2s. 6d.

A Practical Plan for Assimilating the English and American Money as a Step towards a Universal Money. Cr. 8vo. 2s. 6d.

BAGWELL.—**Ireland under the Tudors**, with a Succinct Account of the Earlier History. By RICHARD BAGWELL, M.A. (3 vols.) Vols. I. and II. From the first invasion of the Northmen to the year 1578. 8vo. 32s. Vol. III. 1578-1603. 8vo. 18s.

BAIN (Alexander).—WORKS BY.
Mental and Moral Science. Cr. 8vo. 10s. 6d.
Senses and the Intellect. 8vo. 15s.
Emotions and the Will. 8vo. 15s.
Logic, Deductive, and Inductive. PART I. *Deduction*, 4s. PART II. *Induction*, 6s. 6d.
Practical Essays. Cr. 8vo. 2s.

BAKER.—**By the Western Sea**: a Summer Idyll. By JAMES BAKER, F.R.G.S. Author of 'John Westacott'. Crown 8vo. 3s. 6d.

BAKER (Sir S. W.).—WORKS BY.
Eight Years in Ceylon. With 6 Illustrations. Crown 8vo. 3s. 6d.
The Rifle and the Hound in Ceylon. With 6 Illustrations. Crown 8vo. 3s. 6d.

BALL (The Rt. Hon. J. T.).—WORKS BY.
The Reformed Church of Ireland. (1537-1889). 8vo. 7s. 6d.
Historical Review of the Legislative Systems Operative in Ireland, from the Invasion of Henry the Second to the Union (1172-1800). 8vo. 6s.

BEACONSFIELD (The Earl of).—WORKS BY.
Novels and Tales. The Hughenden Edition. With 2 Portraits and 11 Vignettes. 11 vols. Crown 8vo. 42s.
Endymion.   Henrietta Temple.
Lothair.   Contarini, Fleming, &c.
Coningsby.   Alroy, Ixion, &c.
Tancred. Sybil.   The Young Duke, &c.
Venetia.   Vivian Grey.
Novels and Tales. Cheap Edition. Complete in 11 vols. Crown 8vo. 1s. each, boards; 1s. 6d. each, cloth.

BECKER (Professor).—WORKS BY.
Gallus; or, Roman Scenes in the Time of Augustus. Post 8vo. 7s. 6d.
Charicles; or, Illustrations of the Private Life of the Ancient Greeks. Post 8vo. 7s. 6d.

BELL (Mrs. Hugh).—WORKS BY.
Will o' the Wisp: a Story. Illustrated by E. L. SHUTE. Crown 8vo. 3s. 6d.
Chamber Comedies: a Collection of Plays and Monologues for the Drawing Room. Crown 8vo. 6s.

BLAKE.—**Tables for the Conversion of 5 per Cent. Interest from $\frac{1}{16}$ to 7 per Cent.** By J. BLAKE, of the London Joint Stock Bank, Limited. 8vo. 12s. 6d.

Book (The) of Wedding Days. Arranged on the Plan of a Birthday Book. With 96 Illustrated Borders, Frontispiece, and Title-page by WALTER CRANE; and Quotations for each Day. Compiled and Arranged by K. E. J. REID, MAY ROSS, and MABEL BAMFIELD. 4to. 21s.

BRASSEY (Lady).—WORKS BY.
A Voyage in the 'Sunbeam,' our Home on the Ocean for Eleven Months.
Library Edition. With 8 Maps and Charts, and 118 Illustrations, 8vo. 21s.
Cabinet Edition. With Map and 66 Illustrations, Crown 8vo. 7s. 6d.
'Silver Library' Edition. With 66 Illustrations. Crown 8vo. 3s. 6d.
Popular Edition. With 60 Illustrations, 4to. 6d. sewed, 1s. cloth.
School Edition. With 37 Illustrations, Fcp. 2s. cloth, or 3s. white parchment.
Sunshine and Storm in the East.
Library Edition. With 2 Maps and 114 Illustrations, 8vo. 21s.
Cabinet Edition. With 2 Maps and 114 Illustrations, Crown 8vo. 7s. 6d.
Popular Edition. With 103 Illustrations, 4to. 6d. sewed, 1s. cloth.
In the Trades, the Tropics, and the 'Roaring Forties'.
Cabinet Edition. With Map and 220 Illustrations, Crown 8vo. 7s. 6d.
Popular Edition. With 183 Illustrations, 4to. 6d. sewed, 1s. cloth.
The Last Voyage to India and Australia in the 'Sunbeam'. With Charts and Maps, and 40 Illustrations in Monotone (20 full-page), and nearly 200 Illustrations in the Text from Drawings by R. T. PRITCHETT. 8vo. 21s.
Three Voyages in the 'Sunbeam'. Popular Edition. With 346 Illustrations, 4to. 2s. 6d.

BRAY.—**The Philosophy of Necessity**; or, Law in Mind as in Matter. By CHARLES BRAY. Crown 8vo. 5s.

BRIGHT.—**A History of England.**
By the Rev. J. FRANCK BRIGHT, D.D., Master of University College, Oxford. 4 vols. Crown 8vo.
Period I.—Mediæval Monarchy: The Departure of the Romans to Richard III. From A.D. 449 to 1485. 4s. 6d.
Period II.—Personal Monarchy: Henry VII. to James II. From 1485 to 1688. 5s.
Period III.—Constitutional Monarchy: William and Mary to William IV. From 1689 to 1837. 7s. 6d.
Period IV.—The Growth of Democracy: Victoria. From 1837 to 1880. 6s.

BROKE.—**With Sack and Stock in Alaska.** By GEORGE BROKE, A.C., F.R.G.S. With 2 Maps. Crown 8vo. 5s.

BRYDEN.—**Kloof and Karroo:** Sport, Legend, and Natural History in Cape Colony. By H. A. BRYDEN. With 17 Illustrations. 8vo. 10s. 6d.

BUCKLE.—**History of Civilisation in England and France, Spain and Scotland.** By HENRY THOMAS BUCKLE. 3 vols. Cr. 8vo. 24s.

BULL (Thomas).—*WORKS BY.*
Hints to Mothers on the Management of their Health during the Period of Pregnancy. Fcp. 8vo. 1s. 6d.

The Maternal Management of Children in Health and Disease. Fcp. 8vo. 1s. 6d.

BUTLER (Samuel).—*WORKS BY.*
Op. 1. **Erewhon.** Crown 8vo. 5s.
Op. 2. **The Fair Haven.** A Work in defence of the Miraculous Element in our Lord's Ministry. Crown 8vo. 7s. 6d.
Op. 3. **Life and Habit.** An Essay after a Completer View of Evolution. Crown 8vo. 7s. 6d.
Op. 4. **Evolution, Old and New.** Crown 8vo. 10s. 6d.
Op. 5. **Unconscious Memory.** Crown 8vo. 7s. 6d.
Op. 6. **Alps and Sanctuaries of Piedmont and the Canton Ticino.** Illustrated. Pott 4to. 10s. 6d.
Op. 7. **Selections from Ops. 1-6.** With Remarks on Mr. G. J. ROMANES' 'Mental Evolution in Animals'. Cr. 8vo. 7s. 6d.

BUTLER (Samuel).—*WORKS BY.*—*continued.*
Op. 8. **Luck, or Cunning, as the Main Means of Organic Modification?** Cr. 8vo. 7s. 6d.
Op. 9. **Ex Voto.** An Account of the Sacro Monte or New Jerusalem at Varallo-Sesia. 10s. 6d.
Holbein's 'La Danse'. A Note on a Drawing called 'La Danse'. 3s.

CARLYLE.—**Thomas Carlyle: a** History of His Life. By J. A. FROUDE. 1795-1835, 2 vols. Crown 8vo. 7s. 1834-1881, 2 vols. Crown 8vo. 7s.

CASE.—**Physical Realism:** being an Analytical Philosophy from the Physical Objects of Science to the Physical Data of Sense. By THOMAS CASE, M.A., Fellow and Senior Tutor, C.C.C. 8vo. 15s.

CHETWYND.—**Racing Reminiscences and Experiences of the Turf.** By Sir GEORGE CHETWYND, Bart. 2 vols. 8vo. 21s.

CHILD.—**Church and State under the Tudors.** By GILBERT W. CHILD, M.A., Exeter College, Oxford. 8vo. 15s.

CHISHOLM.—**Handbook of Commercial Geography.** By G. G. CHISHOLM, B.Sc. With 29 Maps. 8vo. 16s.

CHURCH.—**Sir Richard Church, C.B., G.C.H.** Commander-in-Chief of the Greeks in the War of Independence: a Memoir. By STANLEY LANE-POOLE. With 2 Plans. 8vo. 5s.

CLIVE.—**Poems.** By V. (Mrs. ARCHER CLIVE), Author of 'Paul Ferroll'. Including the IX. Poems. Fcp. 8vo. 6s.

CLODD.—**The Story of Creation:** a Plain Account of Evolution. By EDWARD CLODD. With 77 Illustrations. Crown 8vo. 3s. 6d.

CLUTTERBUCK.—**The Skipper in Arctic Seas.** By W. J. CLUTTERBUCK, one of the Authors of 'Three in Norway'. With 39 Illustrations. Cr. 8vo. 10s. 6d.

COLENSO.—**The Pentateuch and Book of Joshua Critically Examined.** By J. W. COLENSO, D.D., late Bishop of Natal. Cr. 8vo. 6s.

COLMORE.—A Living Epitaph. By G. COLMORE, Author of 'A Conspiracy of Silence,' &c. Crown 8vo. 6s.

COMYN.—Atherstone Priory: a Tale. By L. N. COMYN. Cr. 8vo. 2s. 6d.

CONINGTON (John).—WORKS BY.

The Æneid of Virgil. Translated into English Verse. Crown 8vo. 6s.

The Poems of Virgil. Translated into English Prose. Crown 8vo. 6s.

COX.—A General History of Greece, from the Earliest Period to the Death of Alexander the Great; with a sketch of the subsequent History to the Present Time. By the Rev. Sir G. W. COX, Bart., M.A. With 11 Maps and Plans. Crown 8vo. 7s. 6d.

CRAKE (Rev. A. D.).—WORKS BY.

Historical Tales. Crown 8vo. 5 vols. 2s. 6d. each.

Edwy the Fair; or, The First Chronicle of Æscendune.

Alfgar the Dane; or, the Second Chronicle of Æscendune.

The Rival Heirs: being the Third and Last Chronicle of Æscendune.

The House of Walderne. A Tale of the Cloister and the Forest in the Days of the Barons' Wars.

Brian Fitz-Count. A Story of Wallingford Castle and Dorchester Abbey.

History of the Church under the Roman Empire, A.D. 30-476. Crown 8vo. 7s. 6d.

CREIGHTON.—History of the Papacy during the Reformation. By MANDELL CREIGHTON, D.D., LL.D., Bishop of Peterborough. 8vo. Vols. I. and II., 1378-1464, 32s.; Vols. III. and IV., 1464-1518, 24s.

CRUMP (A.).—WORKS BY.

A Short Enquiry into the Formation of Political Opinion, from the reign of the Great Families to the Advent of Democracy. 8vo. 7s. 6d.

An Investigation into the Causes of the Great Fall in Prices which took place coincidently with the Demonetisation of Silver by Germany. 8vo. 6s.

CUDWORTH.—An Introduction to Cudworth's Treatise concerning Eternal and Immutable Morality. With Life of Cudworth, and a few Critical Notes. By W. R. SCOTT, First Senior Moderator in Logics and Ethics, Trinity College, Dublin. Crown 8vo. 3s.

CURZON.—Russia in Central Asia in 1889, and the Anglo-Russian Question. By the Hon. GEORGE N. CURZON, M.P. 8vo. 21s.

DANTE.—La Commedia di Dante. A New Text, carefully Revised with the aid of the most recent Editions and Collations. Small 8vo. 6s.

*_** Fifty Copies (of which Forty-five are for Sale) have been printed on Japanese paper, £1 1s. net.

DAVIDSON (W. L.).—WORKS BY.

The Logic of Definition Explained and Applied. Cr. 8vo. 6s.

Leading and Important English Words Explained and Exemplified. Fcp. 8vo. 3s. 6d.

DELAND (Mrs.).—WORKS BY.

John Ward, Preacher: a Story. Crown 8vo. 2s. boards, 2s. 6d. cloth.

Sidney: a Novel. Crown 8vo. 6s.

The Old Garden, and other Verses. Fcp. 8vo. 5s.

Florida Days. With 12 Full-page Plates (2 Etched and 4 in Colours), and about 50 Illustrations in the Text, by LOUIS K. HARLOW. 8vo. 21s.

DE LA SAUSSAYE.—A Manual of the Science of Religion. By Professor CHANTEPIE DE LA SAUSSAYE. Translated by Mrs. COLYER FERGUSSON (née MAX MÜLLER). Revised by the Author.

DE REDCLIFFE.—The Life of the Right Hon. Stratford Canning: Viscount Stratford De Redcliffe. By STANLEY LANE-POOLE.

Cabinet Edition, abridged, with 3 Portraits, 1 vol. Crown 8vo. 7s. 6d.

DE SALIS (Mrs.).—WORKS BY.

Savouries à la Mode. Fcp. 8vo. 1s. 6d. boards.

Entrées à la Mode. Fcp. 8vo. 1s. 6d. boards.

[Continued on next page.

DE SALIS(Mrs.).— WORKS BY.—cont.

Soups and Dressed Fish à la Mode. Fcp. 8vo. 1s. 6d. boards.

Oysters à la Mode. Fcp. 8vo. 1s. 6d. boards.

Sweets and Supper Dishes à la Mode. Fcp. 8vo. 1s. 6d. boards.

Dressed Vegetables à la Mode. Fcp. 8vo. 1s. 6d. boards.

Dressed Game and Poultry à la Mode. Fcp. 8vo. 1s. 6d. bds.

Drinks à la Mode. Fcp. 8vo. 1s. 6d. boards.

Puddings and Pastry à la Mode. Fcp. 8vo. 1s. 6d. boards.

Cakes and Confections à la Mode. Fcp. 8vo. 1s. 6d. boards.

Tempting Dishes for Small Incomes. Fcp. 8vo. 1s. 6d.

Floral Decorations. Suggestions and Descriptions. Fcap. 8vo. 1s. 6d.

Wrinkles and Notions for every Household. Crown 8vo. 2s. 6d.

DE TOCQUEVILLE.—**Democracy in America.** By ALEXIS DE TOCQUEVILLE. Translated by HENRY REEVE, C.B. 2 vols. Crown 8vo. 16s.

DOWELL.—**A History of Taxation and Taxes in England** from the Earliest Times to the Year 1885. By STEPHEN DOWELL. (4 vols. 8vo.) Vols. I. and II. The History of Taxation, 21s. Vols. III. and IV. The History of Taxes, 21s.

DOYLE (A. Conan).— WORKS BY.

Micah Clarke. A tale of Monmouth's Rebellion. With Frontispiece and Vignette. Crown 8vo. 3s. 6d.

The Captain of the Polestar; and other Tales. Crown 8vo. 6s.

DRANE.—**The History of St. Dominic,** Founder of the Friar Preachers. By AUGUSTA THEODORA DRANE. With 32 Illustrations. 8vo. 15s.

Dublin University Press Series (The): a Series of Works undertaken by the Provost and Senior Fellows of Trinity College, Dublin.

Abbott's (T. K.) Codex Rescriptus Dublinensis of St. Matthew. 4to. 21s.

——— Evangeliorum Versio Antehieronymiana ex Codice Usseriano (Dublinensi). 2 vols. Crown 8vo. 21s.

Dublin University Press Series (The).—*continued.*

Allman's (G. J.) Greek Geometry from Thales to Euclid. 8vo. 10s. 6d.

Burnside (W. S.) and Panton's (A. W.) Theory of Equations. 8vo. 12s. 6d.

Casey's (John) Sequel to Euclid's Elements. Crown 8vo. 3s. 6d.

——— Analytical Geometry of the Conic Sections. Crown 8vo. 7s. 6d.

Davies' (J. F.) Eumenides of Æschylus. With Metrical English Translation. 8vo. 7s.

Dublin Translations into Greek and Latin Verse. Edited by R. Y. Tyrrell. 8vo. 6s.

Graves' (R. P.) Life of Sir William Hamilton. 3 vols. 15s. each.

Griffin (R. W.) on Parabola, Ellipse, and Hyperbola. Crown 8vo. 6s.

Hobart's (W. K.) Medical Language of St. Luke. 8vo. 16s.

Leslie's (T. E. Cliffe) Essays in Political Economy. 8vo. 10s. 6d.

Macalister's (A.) Zoology and Morphology of Vertebrata. 8vo. 10s. 6d.

MacCullagh's (James) Mathematical and other Tracts. 8vo. 15s.

Maguire's (T.) Parmenides of Plato, Text, with Introduction, Analysis, &c. 8vo. 7s. 6d.

Monck's (W. H. S.) Introduction to Logic. Crown 8vo. 5s.

Roberts' (R. A.) Examples on the Analytic Geometry of Plane Conics. Cr. 8vo. 5s.

Southey's (R.) Correspondence with Caroline Bowles. Edited by E. Dowden. 8vo. 14s.

Stubbs' (J. W.) History of the University of Dublin, from its Foundation to the End of the Eighteenth Century. 8vo. 12s. 6d.

Thornhill's (W. J.) The Æneid of Virgil, freely translated into English Blank Verse. Crown 8vo. 7s. 6d.

Tyrrell's (R. Y.) Cicero's Correspondence. Vols. I. II. III. 8vo. each 12s.

——— The Acharnians of Aristophanes, translated into English Verse. Crown 8vo. 1s.

Webb's (T. E.) Goethe's Faust, Translation and Notes. 8vo. 12s. 6d.

——— The Veil of Isis: a Series of Essays on Idealism. 8vo. 10s. 6d.

Wilkins' (G.) The Growth of the Homeric Poems. 8vo. 6s.

Epochs of Modern History.

Edited by C. COLBECK, M.A. 19 vols. Fcp. 8vo. with Maps, 2s. 6d. each.

Church's (Very Rev. R. W.) The Beginning of the Middle Ages. With 3 Maps.

Johnson's (Rev. A. H.) The Normans in Europe. With 3 Maps.

Cox's (Rev. Sir G. W.) The Crusades. With a Map.

Stubbs' (Right Rev. W.) The Early Plantagenets. With 2 Maps.

Warburton's (Rev. W.) Edward the Third. With 3 Maps.

Gairdner's (J.) The Houses of Lancaster and York; with the Conquest and Loss of France. With 5 Maps.

Moberly's (Rev. C. E.) The Early Tudors.

Seebohm's (F.) The Era of the Protestant Revolution. With 4 Maps.

Creighton's (Rev. M.) The Age of Elizabeth. With 5 Maps.

Gardiner's (S. R.) The First Two Stuarts and the Puritan Revolution (1603-1660). With 4 Maps.

——— The Thirty Years' War (1618-1648). With a Map.

Airy's (O.) The English Restoration and Louis XIV. (1648-1678).

Hale's (Rev. E.) The Fall of the Stuarts; and Western Europe (1678-1697). With 11 Maps and Plans.

Morris's (E. E.) The Age of Anne. With 7 Maps and Plans.

——— The Early Hanoverians. With 9 Maps and Plans.

Longman's (F. W.) Frederick the Great and the Seven Years' War. With 2 Maps.

Ludlow's (J. M.) The War of American Independence (1775-1783). With 4 Maps.

Gardiner's (Mrs. S. R.) The French Revolution (1789-1795). With 7 Maps.

McCarthy's (Justin) The Epoch of Reform (1830-1850).

Epochs of Church History.

Edited by MANDELL CREIGHTON, D.D., Bishop of Peterborough. Fcp. 8vo. 2s. 6d. each.

Tucker's (Rev. H. W.) The English Church in other Lands.

Perry's (Rev. G. G.) The History of the Reformation in England.

Brodrick's (Hon. G. C.) A History of the University of Oxford.

Epochs of Church History.—cont.

Mullinger's (J. B.) A History of the University of Cambridge.

Plummer's (A.) The Church of the Early Fathers.

Carr's (Rev. A.) The Church and the Roman Empire.

Wakeman's (H. O.) The Church and the Puritans (1570-1660.)

Overton's (Rev. J. H.) The Evangelical Revival in the Eighteenth Century.

Tozer's (Rev. H. F.) The Church and the Eastern Empire.

Stephen's (Rev. W. R. W.) Hildebrand and his Times.

Hunt's (Rev. W.) The English Church in the Middle Ages.

Balzani's (U.) The Popes and the Hohenstaufen.

Gwatkin's (H. M.) The Arian Controversy.

Ward's (A. W.) The Counter-Reformation.

Poole's (R. L.) Wycliffe and Early Movements of Reform.

Epochs of Ancient History.

Edited by the Rev. Sir G. W. COX, Bart., M.A., and by C. SANKEY, M.A. 10 volumes, Fcp. 8vo. with Maps, 2s. 6d. each.

Beesly's (A. H.) The Gracchi, Marius, and Sulla. With 2 Maps.

Capes' (Rev. W. W.) The Early Roman Empire. From the Assassination of Julius Cæsar to the Assassination of Domitian. With 2 Maps.

——— The Roman Empire of the Second Century, or the Age of the Antonines. With 2 Maps.

Cox's (Rev. Sir G. W.) The Athenian Empire from the Flight of Xerxes to the Fall of Athens. With 5 Maps.

——— The Greeks and the Persians. With 4 Maps.

Curteis's (A. M.) The Rise of the Macedonian Empire. With 8 Maps.

Ihne's (W.) Rome to its Capture by the Gauls. With a Map.

Merivale's (Very Rev. C.) The Roman Triumvirates. With a Map.

Sankey's (C.) The Spartan and Theban Supremacies. With 5 Maps.

Smith's (R. B.) Rome and Carthage, the Punic Wars. With 9 Maps and Plans.

Epochs of American History. Edited by Dr. ALBERT BUSHNELL HART, Assistant Professor of History in Harvard College.

Thwaites's (R. G.) The Colonies (1492-1763). Fcp. 8vo. 3s. 6d. [Ready.

Hart's (A. B.) Formation of the Union (1763-1829). Fcp. 8vo. [In preparation.

Wilson's (W.) Division and Re-union (1829-1889). Fcp. 8vo. [In preparation.

Epochs of English History. Complete in One Volume, with 27 Tables and Pedigrees, and 23 Maps. Fcp. 8vo. 5s.

_{*} For details of Parts *see* Longmans & Co.'s Catalogue of School Books.

EWALD (Heinrich).— WORKS BY.

The Antiquities of Israel. Translated from the German by H. S. SOLLY, M.A. 8vo. 12s. 6d.

The History of Israel. Translated from the German. 8 vols. 8vo. Vols. I. and II. 24s. Vols. III. and IV. 21s. Vol. V. 18s. Vol. VI. 16s. Vol. VII. 21s. Vol. VIII., with Index to the Complete Work, 18s.

FARNELL.—**Greek Lyric Poetry:** a Complete Collection of the Surviving Passages from the Greek Song-Writers. Arranged with Prefatory Articles, Introductory Matter, and Commentary. By GEORGE S. FARNELL, M.A. With 5 Plates. 8vo. 16s.

FARRAR (Ven. Archdeacon).—WORKS BY.

Darkness and Dawn: a Story of the Early Christians. 2 vols. 8vo. [In the Press.

Language and Languages. A Revised Edition of *Chapters on Language and Families of Speech.* Crown 8vo. 6s.

FIRTH.—**Nation Making:** a Story of New Zealand Savageism and Civilisation. By J. C. FIRTH, Author of 'Luck' and 'Our Kin across the Sea'. Crown 8vo. 6s.

FITZWYGRAM. — **Horses and Stables.** By Major-General Sir F. FITZWYGRAM, Bart. With 19 pages of Illustrations. 8vo. 5s.

FORD.—**The Theory and Practice of Archery.** By the late HORACE FORD. New Edition, thoroughly Revised and Re-written by W. BUTT, M.A. With a Preface by C. J. LONGMAN, M.A., F.S.A. 8vo. 14s.

FOUARD.—**The Christ the Son of God:** a Life of our Lord and Saviour Jesus Christ. By the Abbé CONSTANT FOUARD. Translated from the Fifth Edition, with the Author's sanction, by GEORGE F. X. GRIFFITH. With an Introduction by Cardinal MANNING. 2 vols. Crown 8vo. 14s.

FOX. — **The Early History of Charles James Fox.** By the Right Hon. Sir G. O. TREVELYAN, Bart. Library Edition, 8vo. 18s. Cabinet Edition, Crown 8vo. 6s.

FRANCIS.—**A Book on Angling;** or, Treatise on the Art of Fishing in every branch; including full Illustrated List of Salmon Flies. By FRANCIS FRANCIS. With Portrait and Coloured Plates. Crown 8vo. 15s.

FREEMAN.—**The Historical Geography of Europe.** By E. A. FREEMAN. With 65 Maps. 2 vols. 8vo. 31s. 6d.

FROUDE (James A.).—WORKS BY.

The History of England, from the Fall of Wolsey to the Defeat of the Spanish Armada. 12 vols. Crown 8vo. 3s. 6d. each.

Short Studies on Great Subjects. Cabinet Edition, 4 vols. Crown 8vo. 24s. Cheap Edition, 4 vols. Crown 8vo. 3s. 6d. each.

Cæsar: a Sketch. Crown 8vo. 3s. 6d.

The English in Ireland in the Eighteenth Century. 3 vols. Crown 8vo. 18s.

Oceana; or, England and her Colonies. With 9 Illustrations. Crown 8vo. 2s. boards, 2s. 6d. cloth.

The English in the West Indies; or, the Bow of Ulysses. With 9 Illustrations. Crown 8vo. 2s. boards, 2s. 6d. cloth.

The Two Chiefs of Dunboy; an Irish Romance of the Last Century. Crown 8vo. 3s. 6d.

Thomas Carlyle, a History of his Life. 1795 to 1835. 2 vols. Crown 8vo. 7s. 1834 to 1881. 2 vols. Crown 8vo. 7s.

GALLWEY.—**Letters to Young Shooters.** (First Series.) On the Choice and Use of a Gun. By Sir RALPH PAYNE-GALLWEY, Bart. With Illustrations. Crown 8vo. 7s. 6d.

GARDINER (Samuel Rawson).—WORKS BY.

History of England, from the Accession of James I. to the Outbreak of the Civil War, 1603-1642. 10 vols. Crown 8vo. price 6s. each.

A History of the Great Civil War, 1642-1649. (3 vols.) Vol. I. 1642-1644. With 24 Maps. 8vo. 21s. (*out of print*). Vol. II. 1644-1647. With 21 Maps. 8vo. 24s. Vol. III. 1647-1649 (*in the press*).

The Student's History of England. Illustrated under the superintendence of ST. JOHN HOPE, Secretary to the Society of Antiquaries. Vol. I. B.C. 55—A.D. 1509, with 173 Illustrations. Crown 8vo. 4s. Vol. II. 1509-1689, with 96 Illustrations. Crown 8vo. 4s. Vol. III. (1689-1865). Crown 8vo. 4s.

The work will be published in Three Volumes, and also in One Volume complete.

GIBERNE.—WORKS BY.

Ralph Hardcastle's Will. By AGNES GIBERNE. With Frontispiece. Crown 8vo. 5s.

Nigel Browning. Crown 8vo. 5s.

Miss Devereux, Spinster. A Novel. 2 vols. Crown 8vo. 17s.

*GOETHE.—***Faust.** A New Translation chiefly in Blank Verse; with Introduction and Notes. By JAMES ADEY BIRDS. Crown 8vo. 6s.

Faust. The Second Part. A New Translation in Verse. By JAMES ADEY BIRDS. Crown 8vo. 6s.

*GREEN.—***The Works of Thomas Hill Green.** Edited by R. L. NETTLESHIP. (3 vols.) Vols. I. and II. —Philosophical Works. 8vo. 16s. each. Vol. III.—Miscellanies. With Index to the three Volumes and Memoir. 8vo. 21s.

The Witness of God and Faith: Two Lay Sermons. By T. H. GREEN. Fcp. 8vo. 2s.

*GREVILLE.—***A Journal of the Reigns of King George IV., King William IV., and Queen Victoria.** By C. C. F. GREVILLE. Edited by H. REEVE. 8 vols. Crown 8vo. 6s. each.

GWILT.— **An Encyclopædia of Architecture.** By JOSEPH GWILT, F.S.A. Illustrated with more than 1700 Engravings on Wood. 8vo. 52s. 6d.

*HAGGARD.—***Life and its Author:** an Essay in Verse. By ELLA HAGGARD. With a Memoir by H. RIDER HAGGARD, and Portrait. Fcp. 8vo. 3s. 6d.

HAGGARD (H. Rider).—WORKS BY.

She. With 32 Illustrations by M. GREIFFENHAGEN and C. H. M. KERR. Crown 8vo. 3s. 6d.

Allan Quatermain. With 31 Illustrations by C. H. M. KERR. Crown 8vo. 3s. 6d.

Maiwa's Revenge; or, The War of the Little Hand. Crown 8vo. 1s. boards; 1s. 6d. cloth.

Colonel Quaritch, V.C. A Novel. Crown 8vo. 3s. 6d.

Cleopatra: being an Account of the Fall and Vengeance of Harmachis, the Royal Egyptian. With 29 Full-page Illustrations by M. Greiffenhagen and R. Caton Woodville. Crown 8vo. 3s. 6d.

Beatrice. A Novel. Cr. 8vo. 6s.

Eric Brighteyes. With 17 Plates and 34 Illustrations in the Text by LANCELOT SPEED. Crown 8vo. 6s.

HAGGARD and LANG.— **The World's Desire.** By H. RIDER HAGGARD and ANDREW LANG. Crown 8vo. 6s.

HALLIWELL-PHILLIPPS.— **A Calendar of the Halliwell-Phillipps' collection of Shakespearean Rarities formerly preserved at Hollingbury Copse, Brighton.** Second Edition. Enlarged by ERNEST E. BAKER, F.S.A. 8vo. 10s. 6d.

HARRISON.— **Myths of the Odyssey in Art and Literature.** Illustrated with Outline Drawings. By JANE E. HARRISON. 8vo. 18s.

HARRISON.— **The Contemporary History of the French Revolution,** compiled from the 'Annual Register'. By F. BAYFORD HARRISON. Crown 8vo. 3s. 6d.

HARTE (Bret).—WORKS BY.
In the Carquinez Woods. Fcp. 8vo. 1s. boards; 1s. 6d. cloth.
On the Frontier. 16mo. 1s.
By Shore and Sedge. 16mo. 1s.

HARTWIG (Dr.).—WORKS BY.
The Sea and its Living Wonders. With 12 Plates and 303 Woodcuts. 8vo. 10s. 6d.
The Tropical World. With 8 Plates and 172 Woodcuts. 8vo. 10s. 6d.
The Polar World. With 3 Maps, 8 Plates and 85 Woodcuts. 8vo. 10s. 6d.
The Subterranean World. With 3 Maps and 80 Woodcuts. 8vo. 10s. 6d.
The Aerial World. With Map, 8 Plates and 60 Woodcuts. 8vo. 10s. 6d.

HAVELOCK.— Memoirs of Sir Henry Havelock, K.C.B. By John Clark Marshman. Crown 8vo. 3s. 6d.

HEARN (W. Edward).—WORKS BY.
The Government of England: its Structure and its Development. 8vo. 16s.
The Aryan Household: its Structure and its Development. An Introduction to Comparative Jurisprudence. 8vo. 16s.

HISTORIC TOWNS. Edited by E. A. Freeman, D.C.L., and Rev. William Hunt, M.A. With Maps and Plans. Crown 8vo. 3s. 6d. each.
Bristol. By Rev. W. Hunt.
Carlisle. By Rev. Mandell Creighton.
Cinque Ports. By Montagu Burrows.
Colchester. By Rev. E. L. Cutts.
Exeter. By E. A. Freeman.
London. By Rev. W. J. Loftie.
Oxford. By Rev. C. W. Boase.
Winchester. By Rev. G. W. Kitchin, D.D.
New York. By Theodore Roosevelt.
Boston (U.S.). By Henry Cabot Lodge.
York. By Rev. James Raine. [In Preparation.

HODGSON (Shadworth H.).—WORKS BY.
Time and Space: a Metaphysical Essay. 8vo. 16s.
The Theory of Practice: an Ethical Enquiry. 2 vols. 8vo. 24s.
The Philosophy of Reflection: 2 vols. 8vo. 21s.
Outcast Essays and Verse Translations. Essays: The Genius of De Quincey—De Quincey as Political Economist—De Quincey and the Supernatural in English Poetry; with Note on the True Symbol of Christian Union—English Verse. Verse Translations: Nineteen Passages from Lucretius, Horace, Homer, &c. Crown 8vo. 8s. 6d.

HOWITT.—Visits to Remarkable Places, Old Halls, Battle-Fields, Scenes, illustrative of Striking Passages in English History and Poetry. By William Howitt. With 80 Illustrations. Crown 8vo. 3s. 6d.

HULLAH (John).—WORKS BY.
Course of Lectures on the History of Modern Music. 8vo. 8s. 6d.
Course of Lectures on the Transition Period of Musical History. 8vo. 10s. 6d.

HUME.--The Philosophical Works of David Hume. Edited by T. H. Green and T. H. Grose. 4 vols. 8vo. 56s. Or Separately, Essays, 2 vols. 28s. Treatise of Human Nature. 2 vols. 28s.

HUTCHINSON (Horace).—WORKS BY.
Cricketing Saws and Stories. With rectilinear Illustrations by the Author. 16mo. 1s.
Creatures of Circumstance: A Novel. 3 vols. Crown 8vo. 25s. 6d.
Famous Golf Links. By Horace G. Hutchinson, Andrew Lang, H. S. C. Everard, T. Rutherford Clark, &c. With numerous Illustrations by F. P. Hopkins, T. Hodges, H. S. King, and from Photographs. Crown 8vo. 6s.

HUTH.—The Marriage of Near Kin, considered with respect to the Law of Nations, the Result of Experience, and the Teachings of Biology. By Alfred H. Huth. Royal 8vo. 21s.

INGELOW (Jean).—WORKS BY.
Poetical Works. Vols. I. and II. Fcp. 8vo. 12s. Vol. III. Fcp. 8vo. 5s.
Lyrical and other Poems. Selected from the Writings of JEAN INGELOW. Fcp. 8vo. 2s. 6d. cloth plain; 3s. cloth gilt.
Very Young and Quite Another Story: Two Stories. Cr. 8vo. 6s.

JAMESON (Mrs.).—WORKS BY.
Sacred and Legendary Art. With 19 Etchings and 187 Woodcuts. 2 vols. 8vo. 20s. net.
Legends of the Madonna. The Virgin Mary as represented in Sacred and Legendary Art. With 27 Etchings and 165 Woodcuts. 1 vol. 8vo. 10s. net.
Legends of the Monastic Orders. With 11 Etchings and 88 Woodcuts. 1 vol. 8vo. 10s. net.
History of Our Lord. His Types and Precursors. Completed by Lady EASTLAKE. With 31 Etchings and 281 Woodcuts. 2 vols. 8vo. 20s. net.

JEFFERIES (Richard).—WORKS BY.
Field and Hedgerow: last Essays. With Portrait. Crown 8vo. 3s. 6d.
The Story of My Heart: my Autobiography. With Portrait and new Preface by C. J. LONGMAN. Crown 8vo. 3s. 6d.

*JENNINGS.—***Ecclesia Anglicana.** A History of the Church of Christ in England, from the Earliest to the Present Times. By the Rev. ARTHUR CHARLES JENNINGS, M.A. Crown 8vo. 7s. 6d.

JESSOP (G. H.).—WORKS BY.
Judge Lynch: a Tale of the California Vineyards. Crown 8vo. 6s.
Gerald Ffrench's Friends. Cr. 8vo. 6s. A collection of Irish-American character stories.

*JOHNSON.—***The Patentee's Manual;** a Treatise on the Law and Practice of Letters Patent. By J. JOHNSON and J. H. JOHNSON. 8vo. 10s. 6d.

*JORDAN (William Leighton).—***The Standard of Value.** By WILLIAM LEIGHTON JORDAN. 8vo. 6s.

*JUSTINIAN.—***The Institutes of Justinian;** Latin Text, chiefly that of Huschke, with English Introduction, Translation, Notes, and Summary. By THOMAS C. SANDARS, M.A. 8vo. 18s.

KALISCH (M. M.).—WORKS BY.
Bible Studies. Part I. The Prophecies of Balaam. 8vo. 10s. 6d. Part II. The Book of Jonah. 8vo. 10s. 6d.
Commentary on the Old Testament; with a New Translation. Vol. I. Genesis, 8vo. 18s. or adapted for the General Reader, 12s. Vol. II. Exodus, 15s. or adapted for the General Reader, 12s. Vol. III. Leviticus, Part I. 15s. or adapted for the General Reader, 8s. Vol. IV. Leviticus, Part II. 15s. or adapted for the General Reader, 8s.

KANT (Immanuel).—WORKS BY.
Critique of Practical Reason, and other Works on the Theory of Ethics. Translated by T. K. Abbott, B.D. With Memoir. 8vo. 12s. 6d.
Introduction to Logic, and his Essay on the Mistaken Subtilty of the Four Figures. Translated by T. K. Abbott. Notes by S. T. Coleridge. 8vo. 6s.

KENDALL (May).—WORKS BY.
From a Garret. Crown 8vo. 6s.
Dreams to Sell; Poems. Fcp. 8vo. 6s.
'Such is Life': a Novel. Crown 8vo. 6s.

*KENNEDY.—***Pictures in Rhyme.** By ARTHUR CLARK KENNEDY. With 4 Illustrations by MAURICE GREIFFENHAGEN. Crown 8vo. 6s.

*KILLICK.—***Handbook to Mill's System of Logic.** By the Rev. A. H. KILLICK, M.A. Crown 8vo. 3s. 6d.

KNIGHT (E. F.).—WORKS BY.
The Cruise of the 'Alerte'; the Narrative of a Search for Treasure on the Desert Island of Trinidad. With 2 Maps and 23 Illustrations. Crown 8vo. 10s. 6d.
Save Me from my Friends: a Novel. Crown 8vo. 6s.

LADD (George T.).—WORKS BY.
Elements of Physiological Psychology. 8vo. 21s.
Outlines of Physiological Psychology. A Text-book of Mental Science for Academies and Colleges. 8vo. 12s.

LANG (Andrew).—WORKS BY.
Custom and Myth: Studies of Early Usage and Belief. With 15 Illustrations. Crown 8vo. 7s. 6d.
Books and Bookmen. With 2 Coloured Plates and 17 Illustrations. Cr. 8vo. 6s. 6d.
Grass of Parnassus. A Volume of Selected Verses. Fcp. 8vo. 6s.
Ballads of Books. Edited by ANDREW LANG. Fcp. 8vo. 6s.
The Blue Fairy Book. Edited by ANDREW LANG. With 8 Plates and 130 Illustrations in the Text by H. J. Ford and G. P. Jacomb Hood. Cr. 8vo. 6s.
The Red Fairy Book. Edited by ANDREW LANG. With 4 Plates and 96 Illustrations in the Text by H. J. Ford and Lancelot Speed. Crown 8vo. 6s.

LAVIGERIE.—**Cardinal Lavigerie and the African Slave Trade.** 8vo. 14s.

LAYARD.—**Poems.** By NINA F. LAYARD. Crown 8vo. 6s.

LECKY (W. E. H.).—WORKS BY.
History of England in the Eighteenth Century. 8vo. Vols. I. & II. 1700-1760. 36s. Vols. III. & IV. 1760-1784. 36s. Vols. V. & VI. 1784-1793. 36s. Vols. VII. & VIII. 1793-1800. 36s.
The History of European Morals from Augustus to Charlemagne. 2 vols. Crown 8vo. 16s.
History of the Rise and Influence of the Spirit of Rationalism in Europe. 2 vols. Crown 8vo. 16s.

LEES and CLUTTERBUCK.—**B. C. 1887, A Ramble in British Columbia.** By J. A. LEES and W. J. CLUTTERBUCK. With Map and 75 Illustrations. Crown 8vo. 6s.

LEGER.—**A History of Austro-Hungary.** From the Earliest Time to the year 1889. By LOUIS LEGER. With a Preface by E. A. FREEMAN, D.C.L. Crown 8vo. 10s. 6d.

LEWES.—**The History of Philosophy, from Thales to Comte.** By GEORGE HENRY LEWES. 2 vols. 8vo. 32s.

LIDDELL.—**The Memoirs of the Tenth Royal Hussars (Prince of Wales' Own):** Historical and Social. Collected and Arranged by Colonel R. S. LIDDELL, late Commanding Tenth Royal Hussars. With Portraits and Coloured Illustration. Imperial 8vo. 63s.

LLOYD.—**The Science of Agriculture.** By F. J. LLOYD. 8vo. 12s.

LONGMAN (Frederick W.).—WORKS BY.
Chess Openings. Fcp. 8vo. 2s. 6d.
Frederick the Great and the Seven Years' War. Fcp. 8vo. 2s. 6d.

Longman's Magazine. Published Monthly. Price Sixpence.
Vols. 1-16. 8vo. price 5s. each.

Longmans' New Atlas. Political and Physical. For the Use of Schools and Private Persons. Consisting of 40 Quarto and 16 Octavo Maps and Diagrams, and 16 Plates of Views. Edited by GEO. G. CHISHOLM, M.A., B.Sc. Imp. 4to. or Imp. 8vo. 12s. 6d.

LOUDON (J. C.).—WORKS BY.
Encyclopædia of Gardening. With 1000 Woodcuts. 8vo. 21s.
Encyclopædia of Agriculture; the Laying-out, Improvement, and Management of Landed Property. With 1100 Woodcuts. 8vo. 21s.
Encyclopædia of Plants; the Specific Character, &c., of all Plants found in Great Britain. With 12,000 Woodcuts. 8vo. 42s.

LUBBOCK.—**The Origin of Civilisation** and the Primitive Condition of Man. By Sir J. LUBBOCK, Bart., M.P. With 5 Plates and 20 Illustrations in the Text. 8vo. 18s.

LYALL.—**The Autobiography of a Slander.** By EDNA LYALL, Author of 'Donovan,' &c. Fcp. 8vo. 1s. sewed.

LYDE.—**An Introduction to Ancient History:** being a Sketch of the History of Egypt, Mesopotamia, Greece, and Rome. With a Chapter on the Development of the Roman Empire into the Powers of Modern Europe. By LIONEL W. LYDE, M.A. With 3 Coloured Maps. Crown 8vo. 3s.

MACAULAY (Lord).—WORKS OF.

Complete Works of Lord Macaulay:
Library Edition, 8 vols. 8vo. £5 5s.
Cabinet Edition, 16 vols. Post 8vo. £4 16s.

History of England from the Accession of James the Second:
Popular Edition, 2 vols. Crown 8vo. 5s.
Student's Edition, 2 vols. Crown 8vo. 12s.
People's Edition, 4 vols. Crown 8vo. 16s.
Cabinet Edition, 8 vols. Post 8vo. 48s.
Library Edition, 5 vols. 8vo. £4.

Critical and Historical Essays, with Lays of Ancient Rome, in 1 volume:
Popular Edition, Crown 8vo. 2s. 6d.
Authorised Edition, Crown 8vo. 2s. 6d. or 3s. 6d. gilt edges.

Critical and Historical Essays:
Student's Edition, 1 vol. Crown 8vo. 6s.
People's Edition, 2 vols. Crown 8vo. 8s.
Trevelyan Edition, 2 vols. Crown 8vo. 9s.
Cabinet Edition, 4 vols. Post 8vo. 24s.
Library Edition, 3 vols. 8vo. 36s.

Essays which may be had separately price 6d. each sewed, 1s. each cloth:
Addison and Walpole.
Frederick the Great.
Croker's Boswell's Johnson.
Hallam's Constitutional History.
Warren Hastings. (3d. sewed, 6d cloth.)
The Earl of Chatham (Two Essays).
Ranke and Gladstone.
Milton and Machiavelli.
Lord Bacon.
Lord Clive.
Lord Byron, and The Comic Dramatists of the Restoration.

The Essay on Warren Hastings annotated by S. HALES, 1s. 6d.

The Essay on Lord Clive annotated by H. COURTHOPE BOWEN, M.A., 2s. 6d.

Speeches:
People's Edition, Crown 8vo. 3s. 6d.

Lays of Ancient Rome, &c.:
Illustrated by G. Scharf, Fcp. 4to. 10s. 6d.
——————— Bijou Edition, 18mo. 2s. 6d. gilt top.
——————— Popular Edition, Fcp. 4to. 6d. sewed, 1s. cloth.
Illustrated by J. R. Weguelin, Crown 8vo. 3s. 6d. cloth extra, gilt edges.
Cabinet Edition, Post 8vo. 3s. 6d.
Annotated Edition, Fcp. 8vo. 1s. sewed, 1s. 6d. cloth.

MACAULAY (Lord).—WORKS OF.—continued.

Miscellaneous Writings:
People's Edition, 1 vol. Crown 8vo. 4s. 6d.
Library Edition, 2 vols. 8vo. 21s.

Miscellaneous Writings and Speeches:
Popular Edition, 1 vol. Crown 8vo. 2s. 6d.
Student's Edition, in 1 vol. Crown 8vo. 6s.
Cabinet Edition, including Indian Penal Code, Lays of Ancient Rome, and Miscellaneous Poems, 4 vols. Post 8vo. 24s.

Selections from the Writings of Lord Macaulay. Edited, with Occasional Notes, by the Right Hon. Sir G. O. TREVELYAN, Bart. Cr. 8vo. 6s.

The Life and Letters of Lord Macaulay. By the Right Hon. Sir G. O. TREVELYAN, Bart.:
Popular Edition, 1 vol. Crown 8vo. 2s. 6d.
Student's Edition, 1 vol. Crown 8vo. 6s.
Cabinet Edition, 2 vols. Post 8vo. 12s.
Library Edition, 2 vols. 8vo. 36s.

MACDONALD (Geo.).—WORKS BY.

Unspoken Sermons. Three Series. Crown 8vo. 3s. 6d. each.

The Miracles of Our Lord. Crown 8vo. 3s. 6d.

A Book of Strife, in the Form of the Diary of an Old Soul: Poems. 12mo. 6s.

MACFARREN.—Lectures on Harmony. By Sir G. A. MACFARREN. 8vo. 12s.

MACKAIL.—Select Epigrams from the Greek Anthology. Edited, with a Revised Text, Introduction, Translation, and Notes, by J. W. MACKAIL, M.A. 8vo. 16s.

MACLEOD (Henry D.).—WORKS BY.

The Elements of Banking. Crown 8vo. 3s. 6d.

The Theory and Practice of Banking. Vol. I. 8vo. 12s. Vol. II. 14s.

The Theory of Credit. 8vo. Vol. I. 7s. 6d.; Vol. II. Part I. 4s. 6d.; Vol. II. Part II. 10s. 6d.

McCULLOCH.—The Dictionary of Commerce and Commercial Navigation of the late J. R. MCCULLOCH. 8vo. with 11 Maps and 30 Charts, 63s.

MACVINE.—**Sixty-Three Years' Angling,** from the Mountain Streamlet to the Mighty Tay. By JOHN MACVINE. Crown 8vo. 10s. 6d.

MALMESBURY.—**Memoirs of an Ex-Minister.** By the Earl of MALMESBURY. Crown 8vo. 7s. 6d.

MANUALS OF CATHOLIC PHILOSOPHY (*Stonyhurst Series*):

Logic. By RICHARD F. CLARKE, S.J. Crown 8vo. 5s.

First Principles of Knowledge. By JOHN RICKABY, S.J. Crown 8vo. 5s.

Moral Philosophy (Ethics and Natural Law). By JOSEPH RICKABY, S.J. Crown 8vo. 5s.

General Metaphysics. By JOHN RICKABY, S.J. Crown 8vo. 5s.

Psychology. By MICHAEL MAHER, S.J. Crown 8vo. 6s. 6d.

Natural Theology. By BERNARD BOEDDER, S.J. Crown 8vo. 6s. 6d.

A Manual of Political Economy. By C. S. DEVAS, Esq., M.A., Examiner in Political Economy in the Royal University of Ireland. 6s. 6d. [*In preparation.*

MARTINEAU (James).—*WORKS BY.*

Hours of Thought on Sacred Things. Two Volumes of Sermons. 2 vols. Crown 8vo. 7s. 6d. each.

Endeavours after the Christian Life. Discourses. Cr. 8vo. 7s. 6d.

The Seat of Authority in Religion. 8vo. 14s.

Essays, Reviews, and Addresses. 4 vols. Crown 8vo. 7s. 6d. each.

I. Personal: Political.
II. Ecclesiastical: Historical.
III. Theological: Philosophical.
IV. Academical: Religious.

[*In course of publication.*

MASON.—**The Steps of the Sun:** Daily Readings of Prose. Selected by AGNES MASON. 16mo. 3s. 6d.

MATTHEWS (Brander).—*WORKS BY.*

A Family Tree, and other Stories. Crown 8vo. 6s.

Pen and Ink: Papers on Subjects of more or less Importance. Cr. 8vo. 5s.

MAUNDER'S TREASURIES.

Biographical Treasury. With Supplement brought down to 1889, by Rev. JAS. WOOD. Fcp. 8vo. 6s.

Treasury of Natural History; or, Popular Dictionary of Zoology. Fcp. 8vo. with 900 Woodcuts. 6s.

Treasury of Geography, Physical, Historical, Descriptive, and Political. With 7 Maps and 16 Plates. Fcp. 8vo. 9s.

Scientific and Literary Treasury. Fcp. 8vo. 6s.

Historical Treasury: Outlines of Universal History, Separate Histories of all Nations. Fcp. 8vo. 6s.

Treasury of Knowledge and Library of Reference. Comprising an English Dictionary and Grammar, Universal Gazetteer, Classical Dictionary, Chronology, Law Dictionary, &c. Fcp. 8vo. 6s.

The Treasury of Bible Knowledge. By the Rev. J. AYRE, M.A. With 5 Maps, 15 Plates, and 300 Woodcuts. Fcp. 8vo. 6s.

The Treasury of Botany. Edited by J. LINDLEY, F.R.S., and T. MOORE, F.L.S. With 274 Woodcuts and 20 Steel Plates. 2 vols. Fcp. 8vo. 12s.

MAX MÜLLER (F.).—*WORKS BY.*

Selected Essays on Language, Mythology and Religion. 2 vols. Crown 8vo. 16s.

Lectures on the Science of Language. 2 vols. Crown 8vo. 16s.

The Science of Language, Founded on Lectures delivered at the Royal Institution in 1861 and 1863. 2 vols. Crown 8vo. 21s.

Three Lectures on the Science of Language and its Place in General Education, delivered at the Oxford University Extension Meeting, 1889. Crown 8vo. 3s.

Hibbert Lectures on the Origin and Growth of Religion, as illustrated by the Religions of India. Crown 8vo. 7s. 6d.

Introduction to the Science of Religion; Four Lectures delivered at the Royal Institution. Crown 8vo. 7s. 6d.

[*Continued on next page.*

MAX MÜLLER (F.).—WORKS BY.— continued.

Natural Religion. The Gifford Lectures, delivered before the University of Glasgow in 1888. Crown 8vo. 10s. 6d.

Physical Religion. The Gifford Lectures, delivered before the University of Glasgow in 1890. Crown 8vo. 10s. 6d.

The Science of Thought. 8vo. 21s.

Three Introductory Lectures on the Science of Thought. 8vo. 2s. 6d.

Biographies of Words, and the Home of the Aryas. Crown 8vo. 7s. 6d.

A Sanskrit Grammar for Beginners. New and Abridged Edition. By A. A. MACDONELL. Cr. 8vo. 6s.

*MAY.—***The Constitutional History of England** since the Accession of George III. 1760-1870. By the Right Hon. Sir THOMAS ERSKINE MAY, K.C.B. 3 vols. Crown 8vo. 18s.

MEADE (L. T.).—WORKS BY.

The O'Donnells of Inchfawn. With Frontispiece by A. CHASEMORE. Crown 8vo. 6s.

Daddy's Boy. With Illustrations. Crown 8vo. 5s.

Deb and the Duchess. With Illustrations by M. E. EDWARDS. Crown 8vo. 5s.

House of Surprises. With Illustrations by EDITH M. SCANNELL. Cr. 8vo. 3s. 6d.

The Beresford Prize. With Illustrations by M. E. EDWARDS. Crown 8vo. 5s.

MEATH (The Earl of).—WORKS BY.

Social Arrows: Reprinted Articles on various Social Subjects. Cr. 8vo. 5s.

Prosperity or Pauperism? Physical, Industrial, and Technical Training. (Edited by the EARL OF MEATH.) 8vo. 5s.

MELVILLE (G. J. White).—NOVELS BY. Crown 8vo. 1s. each, boards; 1s. 6d. each, cloth.

The Gladiators.   Holmby House.
The Interpreter.   Kate Coventry.
Good for Nothing.   Digby Grand.
The Queen's Maries.   General Bounce.

*MENDELSSOHN.—***The Letters of Felix Mendelssohn.** Translated by Lady WALLACE. 2 vols. Cr. 8vo. 10s.

MERIVALE (The Very Rev. Chas.).—WORKS BY.

History of the Romans under the Empire. Cabinet Edition, 8 vols. Crown 8vo. 48s.

Popular Edition, 8 vols. Crown 8vo. 3s. 6d. each.

The Fall of the Roman Republic: a Short History of the Last Century of the Commonwealth. 12mo. 7s. 6d.

General History of Rome from B.C. 753 to A.D. 476. Cr. 8vo. 7s. 6d.

The Roman Triumvirates. With Maps. Fcp. 8vo. 2s. 6d.

*MILES.—***The Correspondence of William Augustus Miles on the French Revolution, 1789-1817.** Edited by the Rev. CHARLES POPHAM MILES, M.A. 2 vols. 8vo. 32s.

*MILL.—***Analysis of the Phenomena of the Human Mind.** By JAMES MILL. 2 vols. 8vo. 28s.

MILL (John Stuart).—WORKS BY.

Principles of Political Economy. Library Edition, 2 vols. 8vo. 30s. People's Edition, 1 vols. Crown 8vo. 5s.

A System of Logic. Cr. 8vo. 5s.

On Liberty. Crown 8vo. 1s. 4d.

On Representative Government. Crown 8vo. 2s.

Utilitarianism. 8vo. 5s.

Examination of Sir William Hamilton's Philosophy. 8vo. 16s.

Nature, the Utility of Religion, and Theism. Three Essays. 8vo. 5s.

MOLESWORTH (Mrs.).—*WORKS BY.*

Marrying and Giving in Marriage: a Novel. Illustrated. Fcp. 8vo. 2s. 6d.

Silverthorns. Illustrated. Crown 8vo. 5s.

The Palace in the Garden. Illustrated. Crown 8vo. 5s.

The Third Miss St. Quentin. Crown 8vo. 6s.

Neighbours. Illustrated. Crown 8vo. 6s.

The Story of a Spring Morning, &c. Illustrated. Crown 8vo. 5s.

MOON.—**The King's English.** By G. WASHINGTON MOON. Fcp. 8vo. 3s. 6d.

MOORE.—**Dante and his Early Biographers.** By EDWARD MOORE, D.D., Principal of St. Edmund Hall, Oxford. Crown 8vo. 4s. 6d.

MULHALL.—**History of Prices since the Year 1850.** By MICHAEL G. MULHALL. Cr. 8vo. 6s.

MURDOCK.—**The Reconstruction of Europe:** a Sketch of the Diplomatic and Military History of Continental Europe, from the Rise to the Fall of the Second French Empire. By HENRY MURDOCK. Crown 8vo. 9s.

MURRAY.—**A Dangerous Catspaw:** a Story. By DAVID CHRISTIE MURRAY and HENRY MURRAY. Crown 8vo. 2s. 6d.

MURRAY and HERMAN.—**Wild Darrie:** a Story. By CHRISTIE MURRAY and HENRY HERMAN. Crown 8vo. 2s. boards; 2s. 6d. cloth.

NANSEN.—**The First Crossing of Greenland.** By Dr. FRIDTJOF NANSEN. With 5 Maps, 12 Plates, and 150 Illustrations in the Text. 2 vols. 8vo. 36s.

NAPIER.—**The Life of Sir Joseph Napier, Bart., Ex-Lord Chancellor of Ireland.** By ALEX. CHARLES EWALD, F.S.A. With Portrait. 8vo. 15s.

NAPIER.—**The Lectures, Essays, and Letters of the Right Hon. Sir Joseph Napier, Bart.,** late Lord Chancellor of Ireland. 8vo. 12s. 6d.

NESBIT.—**Leaves of Life:** Verses. By E. NESBIT. Crown 8vo. 5s.

NEWMAN.—**The Letters and Correspondence of John Henry Newman** during his Life in the English Church. With a brief Autobiographical Memoir. Arranged and Edited by ANNE MOZLEY. With Portraits. 2 vols. 8vo. 30s. net.

NEWMAN (Cardinal).—*WORKS BY.*

Apologia pro Vitâ Sua. Cabinet Edition, Crown 8vo. 6s. Cheap Edition, Crown 8vo. 3s. 6d.

Sermons to Mixed Congregations. Crown 8vo. 6s.

Sermons on Various Occasions. Crown 8vo. 6s.

The Idea of a University defined and illustrated. Cabinet Edition, Crown 8vo. 7s. Cheap Edition, Crown 8vo. 3s. 6d.

Historical Sketches. 3 vols. Cr. 8vo. 6s. each.

The Arians of the Fourth Century. Cabinet Edition, Crown 8vo. 6s. Cheap Edition, Cr. 8vo. 3s. 6d.

Select Treatises of St. Athanasius in Controversy with the Arians. Freely Translated. 2 vols. Cr. 8vo. 15s.

Discussions and Arguments on Various Subjects. Cabinet Edition, Crown 8vo. 6s. Cheap Edition, Crown 8vo. 3s. 6d.

An Essay on the Development of Christian Doctrine. Cabinet Edition, Crown 8vo. 6s. Cheap Edition, Crown 8vo. 3s. 6d.

Certain Difficulties felt by Anglicans in Catholic Teaching Considered. Cabinet Edition, Vol. I., Crown 8vo. 7s. 6d.; Vol. II., Cr. 8vo. 5s. 6d. Cheap Edition, 2 vols. Cr. 8vo. 3s. 6d. each.

[Continued on next page.

*NEWMAN (Cardinal).—WORKS BY.
—continued.*

The Via Media of the Anglican Church, illustrated in Lectures, &c. 2 vols. Crown 8vo. 6s. each.

Essays, Critical and Historical. Cabinet Edition, 2 vols. Crown 8vo. 12s. Cheap Edition, 2 vols. Crown 8vo. 7s.

Essays on Biblical and on Ecclesiastical Miracles. Cabinet Edition, Crown 8vo. 6s. Cheap Edition, Crown 8vo. 3s. 6d.

Tracts. 1. Dissertatiunculæ. 2. On the Text of the Seven Epistles of St. Ignatius. 3. Doctrinal Causes of Arianism. 4. Apollinarianism. 5. St. Cyril's Formula. 6. Ordo de Tempore. 7. Douay Version of Scripture. Crown 8vo. 8s.

An Essay in Aid of a Grammar of Assent. Cabinet Edition, Crown 8vo. 7s. 6d. Cheap Edition, Crown 8vo. 3s. 6d.

Present Position of Catholics in England. Crown 8vo. 7s. 6d.

Callista: a Tale of the Third Century. Cabinet Edition, Crown 8vo. 6s. Cheap Edition, Crown 8vo. 3s. 6d.

Loss and Gain: a Tale. Cabinet Edition, Crown 8vo. 6s. Cheap Edition, Crown 8vo. 3s. 6d.

The Dream of Gerontius. 16mo. 6d. sewed, 1s. cloth.

Verses on Various Occasions. Cabinet Edition, Crown 8vo. 6s. Cheap Edition, Crown 8vo. 3s. 6d.

⁎⁎ For Cardinal Newman's other Works see Messrs. Longmans & Co.'s *Catalogue of Theological Works.*

NORRIS.—**Mrs. Fenton:** a Sketch. By W. E. NORRIS. Crown 8vo. 6s.

NORTON (Charles L.).—WORKS BY.

Political Americanisms: a Glossary of Terms and Phrases Current at Different Periods in American Politics. Fcp. 8vo. 2s. 6d.

A Handbook of Florida. With 49 Maps and Plans. Fcp. 8vo. 5s.

O'BRIEN.—**When we were Boys:** a Novel. By WILLIAM O'BRIEN, M.P. Cabinet Edition, Crown 8vo. 6s. Cheap Edition, Crown 8vo. 2s. 6d.

OLIPHANT (Mrs.).—NOVELS BY.

Madam. Cr. 8vo. 1s. bds.; 1s. 6d. cl.

In Trust. Cr. 8vo. 1s. bds.; 1s. 6d. cl.

Lady Car: the Sequel of a Life. Crown 8vo. 2s. 6d.

OMAN.—**A History of Greece from the Earliest Times to the Macedonian Conquest.** By C. W. C. OMAN, M.A., F.S.A. With Maps and Plans. Crown 8vo. 4s. 6d.

O'REILLY.—**Hurstleigh Dene:** a Tale. By Mrs. O'REILLY. Illustrated by M. ELLEN EDWARDS. Cr. 8vo. 5s.

PAUL.—**Principles of the History of Language.** By HERMANN PAUL. Translated by H. A. STRONG. 8vo. 10s. 6d.

PAYN (James).—NOVELS BY.

The Luck of the Darrells. Cr. 8vo. 1s. boards; 1s. 6d. cloth.

Thicker than Water. Crown 8vo. 1s. boards; 1s. 6d. cloth.

PERRING (Sir Philip).—WORKS BY.

Hard Knots in Shakespeare. 8vo. 7s. 6d.

The 'Works and Days' of Moses. Crown 8vo. 3s. 6d.

PHILLIPPS-WOLLEY.—**Snap:** a Legend of the Lone Mountain. By C. PHILLIPPS-WOLLEY. With 13 Illustrations by H. G. WILLINK. Cr. 8vo. 6s.

POLE.—**The Theory of the Modern Scientific Game of Whist.** By W. POLE, F.R.S. Fcp. 8vo. 2s. 6d.

POLLOCK.—**The Seal of Fate:** a Novel. By Lady POLLOCK and W. H. POLLOCK. Crown 8vo. 6s.

POOLE.—**Cookery for the Diabetic.** By W. H. and Mrs. POOLE. With Preface by Dr. PAVY. Fcp. 8vo. 2s. 6d.

PRENDERGAST.—**Ireland, from the Restoration to the Revolution,** 1660-1690. By JOHN P. PRENDERGAST. 8vo. 5s.

PRINSEP.—Virginie : a Tale of One Hundred Years Ago. By VAL PRINSEP, A.R.A. 3 vols. Crown 8vo. 25s. 6d.

PROCTOR (R. A.).—WORKS BY.

Old and New Astronomy. 12 Parts, 2s. 6d. each. Supplementary Section, 1s. Complete in 1 vol. 4to. 36s. [In course of publication.

The Orbs Around Us ; a Series of Essays on the Moon and Planets, Meteors and Comets. With Chart and Diagrams. Crown 8vo. 5s.

Other Worlds than Ours ; The Plurality of Worlds Studied under the Light of Recent Scientific Researches. With 14 Illustrations. Crown 8vo. 5s.

The Moon ; her Motions, Aspects Scenery, and Physical Condition. With Plates, Charts, Woodcuts, &c. Cr. 8vo. 5s.

Universe of Stars ; Presenting Researches into and New Views respecting the Constitution of the Heavens. With 22 Charts and 22 Diagrams. 8vo. 10s. 6d.

Larger Star Atlas for the Library, in 12 Circular Maps, with Introduction and 2 Index Pages. Folio, 15s. or Maps only, 12s. 6d.

The Student's Atlas. In Twelve Circular Maps on a Uniform Projection and one Scale. 8vo. 5s.

New Star Atlas for the Library, the School, and the Observatory, in 12 Circular Maps. Crown 8vo. 5s.

Light Science for Leisure Hours. Familiar Essays on Scientific Subjects. 3 vols. Crown 8vo. 5s. each.

Chance and Luck ; a Discussion of the Laws of Luck, Coincidences, Wagers, Lotteries, and the Fallacies of Gambling, &c. Crown 8vo. 2s. boards ; 2s. 6d. cloth.

Studies of Venus-Transits. With 7 Diagrams and 10 Plates. 8vo. 5s.

How to Play Whist : with the Laws and Etiquette of Whist. Crown 8vo. 3s. 6d.

Home Whist : an Easy Guide to Correct Play. 16mo. 1s.

The Stars in their Seasons. An Easy Guide to a Knowledge of the Star Groups, in 12 Maps. Roy. 8vo. 5s.

PROCTOR (R. A.).—WORKS BY.—continued.

Star Primer. Showing the Starry Sky Week by Week, in 24 Hourly Maps. Crown 4to. 2s. 6d.

The Seasons pictured in 48 Sun-Views of the Earth, and 24 Zodiacal Maps, &c. Demy 4to. 5s.

Strength and Happiness. With 9 Illustrations. Crown 8vo. 5s.

Strength : How to get Strong and keep Strong, with Chapters on Rowing and Swimming, Fat, Age, and the Waist. With 9 Illustrations. Crown 8vo. 2s.

Rough Ways Made Smooth. Familiar Essays on Scientific Subjects. Crown 8vo. 5s.

Our Place Among Infinities. A Series of Essays contrasting our Little Abode in Space and Time with the Infinities around us. Crown 8vo. 5s.

The Expanse of Heaven. Essays on the Wonders of the Firmament. Cr. 8vo. 5s.

The Great Pyramid, Observatory, Tomb, and Temple. With Illustrations. Crown 8vo. 5s.

Pleasant Ways in Science. Cr. 8vo. 5s.

Myths and Marvels of Astronomy. Crown 8vo. 5s.

Nature Studies. By GRANT ALLEN, A. WILSON, T. FOSTER, E. CLODD, and R. A. PROCTOR. Crown 8vo. 5s.

Leisure Readings. By E. CLODD, A. WILSON, T. FOSTER, A. C. RANYARD, and R. A. PROCTOR. Crown 8vo. 5s.

PRYCE.—The Ancient British Church : an Historical Essay. By JOHN PRYCE, M.A. Crown 8vo. 6s.

RANSOME.—The Rise of Constitutional Government in England : being a Series of Twenty Lectures on the History of the English Constitution delivered to a Popular Audience. By CYRIL RANSOME, M.A. Crown 8vo. 6s.

RAWLINSON.—The History of Phœnicia. By GEORGE RAWLINSON, M.A., Canon of Canterbury, &c. With numerous Illustrations. 8vo. 24s.

READER.—**Echoes of Thought:** a Medley of Verse. By EMILY E. READER. Fcp. 8vo. 5s. cloth, gilt top.

RENDLE and NORMAN.—**The Inns of Old Southwark,** and their Associations. By WILLIAM RENDLE, F.R.C.S., and PHILIP NORMAN, F.S.A. With numerous Illustrations. Roy. 8vo. 28s.

RIBOT.—**The Psychology of Attention.** By TH. RIBOT. Crown 8vo. 3s.

RICH.—**A Dictionary of Roman and Greek Antiquities.** With 2000 Woodcuts. By A. RICH. Crown 8vo. 7s. 6d.

RICHARDSON.—**National Health.** Abridged from 'The Health of Nations'. A Review of the Works of Sir Edwin Chadwick, K.C.B. By Dr. B. W. RICHARDSON. Crown, 4s. 6d.

RILEY.—**Athos;** or, the Mountain of the Monks. By ATHELSTAN RILEY, M.A., F.R.G.S. With Map and 29 Illustrations. 8vo. 21s.

RILEY.—**Old-Fashioned Roses:** Poems. By JAMES WHITCOMB RILEY. 12mo. 5s.

ROBERTS.—**Greek the Language of Christ and His Apostles.** By ALEXANDER ROBERTS, D.D. 8vo. 18s.

ROGET.—**A History of the 'Old Water-Colour' Society** (now the Royal Society of Painters in Water-Colours). With Biographical Notices of its Older and all its Deceased Members and Associates. By JOHN LEWIS ROGET, M.A. 2 vols. Royal 8vo. 42s.

ROGET.—**Thesaurus of English Words and Phrases.** Classified and Arranged so as to facilitate the Expression of Ideas. By PETER M. ROGET. Crown 8vo. 10s. 6d.

RONALDS.—**The Fly-Fisher's Entomology.** By ALFRED RONALDS. With 20 Coloured Plates. 8vo. 14s.

ROSSETTI.—**A Shadow of Dante:** being an Essay towards studying Himself, his World, and his Pilgrimage. By MARIA FRANCESCA ROSSETTI. With Illustrations. Crown 8vo. 10s. 6d.

RUSSELL.—**A Life of Lord John Russell (Earl Russell, K.G.).** By SPENCER WALPOLE. With 2 Portraits. 2 vols. 8vo. 36s. Cabinet Edition, 2 vols. Crown 8vo. 12s.

SEEBOHM (Frederic).—*WORKS BY.*

The Oxford Reformers—John Colet, Erasmus, and Thomas More; a History of their Fellow-Work. 8vo. 14s.

The English Village Community Examined in its Relations to the Manorial and Tribal Systems, &c. 13 Maps and Plates. 8vo. 16s.

The Era of the Protestant Revolution. With Map. Fcp. 8vo. 2s. 6d.

SEWELL.—**Stories and Tales.** By ELIZABETH M. SEWELL. Crown 8vo. 1s. 6d. each, cloth plain; 2s. 6d. each, cloth extra, gilt edges:—

Amy Herbert.	Laneton Parsonage.
The Earl's Daughter.	Ursula.
The Experience of Life.	Gertrude.
A Glimpse of the World.	Ivors.
Cleve Hall.	Home Life.
Katharine Ashton.	After Life.
Margaret Percival.	

SHAKESPEARE. — **Bowdler's Family Shakespeare.** 1 Vol. 8vo. With 36 Woodcuts, 14s. or in 6 vols. Fcp. 8vo. 21s.

Outline of the Life of Shakespeare. By J. O. HALLIWELL-PHILLIPPS. 2 vols. Royal 8vo. £1 1s.

A Calendar of the Halliwell-Phillipps' Collection of Shakespearean Rarities Formerly Preserved at Hollingbury Copse, Brighton. Enlarged by ERNEST E. BAKER, F.S.A. 8vo. 10s. 6d.

Shakespeare's True Life. By JAMES WALTER. With 500 Illustrations. Imp. 8vo. 21s.

The Shakespeare Birthday Book. By MARY F. DUNBAR. 32mo. 1s. 6d. cloth. With Photographs, 32mo. 5s. Drawing-Room Edition, with Photographs, Fcp. 8vo. 10s. 6d.

SHORT.—**Sketch of the History of the Church of England** to the Revolution of 1688. By T. V. SHORT, D.D. Crown 8vo. 7s. 6d.

Silver Library (The).—Crown 8vo. price 3s. 6d. each volume.

Eight Years in Ceylon. By Sir S. W. Baker. With 6 Illustrations.

Rifle and Hound in Ceylon. By Sir S. W. Baker. With 6 Illustrations.

A Voyage in the 'Sunbeam'. With 66 Illustrations. By Lady Brassey.

Story of Creation: a Plain Account of Evolution. By Edward Clodd. With 77 Illustrations.

Micah Clarke. A Tale of Monmouth's Rebellion. By A. Conan Doyle.

Cæsar: a Sketch. By James A. Froude.

Short Studies on Great Subjects. By James A. Froude. 4 vols.

The Two Chiefs of Dunboy: an Irish Romance of the Last Century. By James A. Froude.

Thomas Carlyle: a History of his Life. By J. A. Froude. 1795-1835. 2 vols. 1834-1818. 2 vols.

Life of the Duke of Wellington. By the Rev. G. R. Gleig. With Portrait.

Allan Quatermain. By H. Rider Haggard. With 20 Illustrations.

Cleopatra. By H. Rider Haggard. With 29 Full-page Illustrations.

Colonel Quaritch, V.C.: a Tale of Country Life. By H. Rider Haggard.

She: A History of Adventure. By H. Rider Haggard. 32 Illustrations.

Visits to Remarkable Places. By Wm. Howitt. 80 Illustrations.

Field and Hedgerow. Last Essays of Richard Jefferies. With Portrait.

The Story of My Heart: My Autobiography. By Richard Jefferies.

The Elements of Banking. By Henry D. Macleod.

Memoirs of Major-General Sir Henry Havelock. By J. Clark Marshman.

History of the Romans under the Empire. By the Very Rev. Charles Merivale. 8 vols.

An Essay in Aid of a Grammar of Assent. By Cardinal Newman.

An Essay on the Development of Christian Doctrine. By Cardinal Newman.

Apologia Pro Vitâ Sua. By Cardinal Newman.

Callista: a Tale of the Third Century. By Cardinal Newman.

Silver Library (The).—*continued.*

Certain Difficulties felt by Anglicans in Catholic Teaching Considered. By Cardinal Newman. 2 vols.

Discussions and Arguments on Various Subjects. By Cardinal Newman.

Essays, Critical, and Historical. By Cardinal Newman. 2 vols.

Essays on Biblical and Ecclesiastical Miracles. By Cardinal Newman.

Loss and Gain: a Tale. By Cardinal Newman.

Parochial and Plain Sermons. By Cardinal Newman. 8 vols.

Selection, adapted to the Seasons of the Ecclesiastical Year, from the 'Parochial and Plain Sermons'. By Cardinal Newman.

The Arians of the Fourth Century. By Cardinal Newman.

The Idea of a University defined and Illustrated. By Cardinal Newman.

Verses on Various Occasions. By Cardinal Newman.

Familiar History of Birds. By Edward Stanley, D.D. 160 Illustrations.

Out of Doors. Original Articles on Practical Natural History. By the Rev. J. G. Wood. 11 Illustrations.

Petland Revisited. By the Rev. J. G. Wood. With 33 Illustrations.

Strange Dwellings: a Description of the Habitations of Animals. By the Rev. J. G. Wood. With 60 Illustrations.

SMITH (Gregory).—**Fra Angelico,** and other Short Poems. By GREGORY SMITH. Crown 8vo. 4s. 6d.

SMITH (R. Bosworth).—**Carthage and the Carthagenians.** By R. BOSWORTH SMITH, M.A. Maps, Plans, &c. Crown 8vo. 6s.

Sophocles. Translated into English Verse. By ROBERT WHITELAW, M.A., Assistant-Master in Rugby School; late Fellow of Trinity College, Cambridge. Crown 8vo. 8s. 6d.

STANLEY.—**A Familiar History of Birds.** By E. STANLEY, D.D. With 160 Woodcuts. Crown 8vo. 3s. 6d.

STEEL (J. H.).—WORKS BY.

A Treatise on the Diseases of the Dog; being a Manual of Canine Pathology. Especially adapted for the Use of Veterinary Practitioners and Students. 88 Illustrations. 8vo. 10s. 6d.

A Treatise on the Diseases of the Ox; being a Manual of Bovine Pathology. Especially adapted for the use of Veterinary Practitioners and Students. 2 Plates and 117 Woodcuts. 8vo. 15s.

A Treatise on the Diseases of the Sheep; being a Manual of Ovine Pathology. Especially adapted for the use of Veterinary Practitioners and Students. With Coloured Plate and 99 Woodcuts. 8vo. 12s.

STEPHEN.—**Essays in Ecclesiastical Biography.** By the Right Hon. Sir J. STEPHEN. Crown 8vo. 7s. 6d.

STEPHENS.—**A History of the French Revolution.** By H. MORSE STEPHENS, Balliol College, Oxford. 3 vols. 8vo. Vol. I. 18s. *Ready.* Vol. II. *in the press.*

STEVENSON (Robt. Louis).—WORKS BY.

A Child's Garden of Verses. Small Fcp. 8vo. 5s.

The Dynamiter. Fcp. 8vo. 1s. sewed; 1s. 6d. cloth.

Strange Case of Dr. Jekyll and Mr. Hyde. Fcp. 8vo. 1s. swd.; 1s. 6d. cloth.

STEVENSON and OSBOURNE.—**The Wrong Box.** By ROBERT LOUIS STEVENSON and LLOYD OSBOURNE. Crown 8vo. 5s.

STOCK.—**Deductive Logic.** By ST. GEORGE STOCK. Fcp. 8vo. 3s. 6d.

'*STONEHENGE*'.—**The Dog in Health and Disease.** By 'STONEHENGE'. With 84 Wood Engravings. Square Crown 8vo. 7s. 6d.

STRONG, LOGEMAN, and WHEELER.—**Introduction to the Study of the History of Language.** By HERBERT A. STRONG, M.A., LL.D.; WILLEM S. LOGEMAN; and BENJAMIN IDE WHEELER. 8vo. 10s. 6d.

SULLY (James).—WORKS BY.

Outlines of Psychology, with Special Reference to the Theory of Education. 8vo. 12s. 6d.

The Teacher's Handbook of Psychology, on the Basis of 'Outlines of Psychology'. Crown 8vo. 6s. 6d.

Supernatural Religion; an Inquiry into the Reality of Divine Revelation. 3 vols. 8vo. 36s.

Reply (A) to Dr. Lightfoot's Essays. By the Author of 'Supernatural Religion'. 8vo. 6s.

SWINBURNE.—**Picture Logic**; an Attempt to Popularise the Science of Reasoning. By A. J. SWINBURNE, B.A. Post 8vo. 5s.

SYMES (James).—WORKS BY.

Prelude to Modern History: being a Brief Sketch of the World's History from the Third to the Ninth Century. With 5 Maps. Crown 8vo. 2s. 6d.

A Companion to School Histories of England; being a Series of Short Essays on the most Important Movements, Social, Literary, and Political, in English History. Crown 8vo. 2s. 6d.

Political Economy: a Short Text-Book of Political Economy. With Problems for Solution, and Hints for Supplementary Reading. Crown 8vo. 2s. 6d.

TAYLOR.—**A Student's Manual of the History of India,** from the Earliest Period to the Present Time. By Colonel MEADOWS TAYLOR, C.S.I., &c. Crown 8vo. 7s. 6d.

THOMPSON (D. Greenleaf).—WORKS BY.

The Problem of Evil: an Introduction to the Practical Sciences. 8vo. 10s. 6d.

A System of Psychology. 2 vols. 8vo. 36s.

The Religious Sentiments of the Human Mind. 8vo. 7s. 6d.

Social Progress: an Essay. 8vo. 7s. 6d.

[*Continued on next page.*

THOMPSON (D. Greenleaf).—WORKS BY.—*continued.*

The Philosophy of Fiction in Literature: an Essay. Cr. 8vo. 6s.

Three in Norway. By Two of Them. With a Map and 59 Illustrations. Cr. 8vo. 2s. boards; 2s. 6d. cloth.

TOYNBEE.—Lectures on the Industrial Revolution of the 18th Century in England. By the late Arnold Toynbee, Tutor of Balliol College, Oxford. Together with a Short Memoir by B. Jowett, Master of Balliol College, Oxford. 8vo. 10s. 6d.

TREVELYAN (Sir G. O., Bart.).—WORKS BY.

The Life and Letters of Lord Macaulay.

Popular Edition, Crown 8vo. 2s. 6d.
Student's Edition, Crown 8vo. 6s.
Cabinet Edition, 2 vols. Cr. 8vo. 12s.
Library Edition, 2 vols. 8vo. 36s.

The Early History of Charles James Fox. Library Edition, 8vo. 18s. Cabinet Edition, Cr. 8vo. 6s.

TROLLOPE (Anthony).—NOVELS BY.

The Warden. Crown 8vo. 1s. boards; 1s. 6d. cloth.

Barchester Towers. Crown 8vo. 1s. boards; 1s. 6d. cloth.

VIRGIL.—Publi Vergili Maronis Bucolica, Georgica, Æneis; The Works of Virgil, Latin Text, with English Commentary and Index. By B. H. Kennedy, D.D. Cr. 8vo. 10s. 6d.

The Æneid of Virgil. Translated into English Verse. By John Conington, M.A. Crown 8vo. 6s.

The Poems of Virgil. Translated into English Prose. By John Conington, M.A. Crown 8vo. 6s.

The Eclogues and Georgics of Virgil. Translated from the Latin by J. W. Mackail, M.A., Fellow of Balliol College, Oxford. Printed on Dutch Hand-made Paper. Royal 16mo. 5s.

WAKEMAN and HASSALL.—Essays Introductory to the Study of English Constitutional History. By Resident Members of the University of Oxford. Edited by Henry Offley Wakeman, M.A., and Arthur Hassall, M.A. Crown 8vo. 6s.

WALKER.—The Correct Card; or How to Play at Whist; a Whist Catechism. By Major A. Campbell-Walker, F.R.G.S. Fcp. 8vo. 2s. 6d.

WALPOLE.—History of England from the Conclusion of the Great War in 1815 to 1858. By Spencer Walpole. Library Edition. 5 vols. 8vo. £4 10s. Cabinet Edition. 6 vols. Crown 8vo. 6s. each.

WELLINGTON.—Life of the Duke of Wellington. By the Rev. G. R. Gleig, M.A. Crown 8vo. 3s. 6d.

WELLS.—Recent Economic Changes and their Effect on the Production and Distribution of Wealth and the Well-being of Society. By David A. Wells, LL.D., D.C.L., late United States Special Commissioner of Revenue, &c. Crown 8vo. 10s. 6d.

WENDT.—Papers on Maritime Legislation, with a Translation of the German Mercantile Laws relating to Maritime Commerce. By Ernest Emil Wendt, D.C.L. Royal 8vo. £1 11s. 6d.

WEYMAN.—The House of the Wolf: a Romance. By Stanley J. Weyman. Crown 8vo. 6s.

WHATELY (E. Jane).—WORKS BY.
English Synonyms. Edited by R. Whately, D.D. Fcp. 8vo. 3s.

Life and Correspondence of Richard Whately, D.D., late Archbishop of Dublin. With Portrait. Crown 8vo. 10s. 6d.

WHATELY (Archbishop).—WORKS BY.

Elements of Logic. Crown 8vo. 4s. 6d.

Elements of Rhetoric. Crown 8vo. 4s. 6d.

Lessons on Reasoning. Fcp. 8vo. 1s. 6d.

Bacon's Essays, with Annotations. 8vo. 10s. 6d.

Whist in Diagrams: a Supplement to American Whist, Illustrated; being a Series of Hands played through, Illustrating the American leads, the new play, the forms of Finesse, and celebrated coups of Masters. With Explanation and Analysis. By G. W. P. Fcp. 8vo. 6s. 6d.

WILCOCKS.—**The Sea Fisherman**, Comprising the Chief Methods of Hook and Line Fishing in the British and other Seas, and Remarks on Nets, Boats, and Boating. By J. C. WILCOCKS. Profusely Illustrated. Crown 8vo. 6s.

WILLICH.—**Popular Tables** for giving Information for ascertaining the value of Lifehold, Leasehold, and Church Property, the Public Funds, &c. By CHARLES M. WILLICH. Edited by H. BENCE JONES. Crown 8vo. 10s. 6d.

WILLOUGHBY.—**East Africa and its Big Game.** The Narrative of a Sporting Trip from Zanzibar to the Borders of the Masai. By Capt. Sir JOHN C. WILLOUGHBY, Bart. Illustrated by G. D. Giles and Mrs. Gordon Hake. Royal 8vo. 21s.

WITT (Prof.).—WORKS BY. Translated by FRANCES YOUNGHUSBAND.

The Trojan War. Crown 8vo. 2s.

Myths of Hellas; or, Greek Tales. Crown 8vo. 3s. 6d.

The Wanderings of Ulysses. Crown 8vo. 3s. 6d.

The Retreat of the Ten Thousand; being the story of Xenophon's 'Anabasis'. With Illustrations. Crown 8vo. 3s. 6d.

WOLFF (Henry W.).—WORKS BY.

Rambles in the Black Forest. Crown 8vo. 7s. 6d.

The Watering Places of the Vosges. Crown 8vo. 4s. 6d.

WOOD (Rev. J. G.).—WORKS BY.

Homes Without Hands; a Description of the Habitations of Animals, classed according to the Principle of Construction. With 140 Illustrations. 8vo. 10s. 6d.

WOOD (Rev. J. G.).—WORKS BY.—continued.

Insects at Home; a Popular Account of British Insects, their Structure, Habits, and Transformations. With 700 Illustrations. 8vo. 10s. 6d.

Insects Abroad; a Popular Account of Foreign Insects, their Structure, Habits, and Transformations. With 600 Illustrations. 8vo. 10s. 6d.

Bible Animals; a Description of every Living Creature mentioned in the Scriptures. With 112 Illustrations. 8vo. 10s. 6d.

Strange Dwellings; a Description of the Habitations of Animals, abridged from 'Homes without Hands'. With 60 Illustrations. Crown 8vo. 3s. 6d.

Out of Doors; a Selection of Original Articles on Practical Natural History. With 11 Illustrations. Crown 8vo. 3s. 6d.

Petland Revisited. With 33 Illustrations. Crown 8vo. 3s. 6d.

YOUATT (William).—WORKS BY.

The Horse. Revised and enlarged. 8vo. Woodcuts, 7s. 6d.

The Dog. Revised and enlarged. 8vo. Woodcuts, 6s.

ZELLER (Dr. E.).—WORKS BY.

History of Eclecticism in Greek Philosophy. Translated by SARAH F. ALLEYNE. Cr. 8vo. 10s. 6d.

The Stoics, Epicureans, and Sceptics. Translated by the Rev. O. J. REICHEL, M.A. Crown 8vo. 15s.

Socrates and the Socratic Schools. Translated by the Rev. O. J. REICHEL, M.A. Cr. 8vo. 10s. 6d.

Plato and the Older Academy. Translated by SARAH F. ALLEYNE and ALFRED GOODWIN, B.A. Crown 8vo. 18s.

The Pre-Socratic Schools: a History of Greek Philosophy from the Earliest Period to the time of Socrates. Translated by SARAH F. ALLEYNE. 2 vols. Crown 8vo. 30s.

Outlines of the History of Greek Philosophy. Translated by SARAH F. ALLEYNE and EVELYN ABBOTT. Crown 8vo. 10s. 6d.

www.ingramcontent.com/pod-product-compliance
Lightning Source LLC
Chambersburg PA
CBHW020906230426
43666CB00008B/1329